THE PLANET FACTORY

Also available in the Bloomsbury Sigma series:

THE PLANET FACTORY

EXOPLANETS AND THE SEARCH
FOR A SECOND EARTH

Elizabeth Tasker

BLOOMSBURY
sigma

Bloomsbury Sigma
An imprint of Bloomsbury Publishing Plc

50 Bedford Square
London
WC1B 3DP
UK

1385 Broadway
New York
NY 10018
USA

www.bloomsbury.com

BLOOMSBURY and the Diana logo are trademarks of Bloomsbury Publishing Plc

First published 2017

British Library Cataloguing-in-Publication Data
A catalogue record for this book is available from the British Library.

Library of Congress Cataloguing-in-Publication data has been applied for.

ISBN (hardback) 978-1-4729-1772-0
ISBN (trade paperback) 978-1-4729-1773-7
ISBN (ebook) 978-1-4729-1775-1

2 4 6 8 10 9 7 5 3 1

Typeset by Deanta Global Publishing Services, Chennai, India
Printed and bound in Great Britain by CPI Group (UK) Ltd. Croydon, CR0 4YY

Illustrations by Elizabeth Tasker

Bloomsbury Sigma, Book Twenty-six

To find out more about our authors and books visit www.bloomsbury.com.
Here you will find extracts, author interviews, details of forthcoming events
and the option to sign up for our newsletters.

To my parents, with whom I was furious from the age of eight for not noticing my initials would be E.T.

... I admit it paid off.

Contents

Contents

Preface

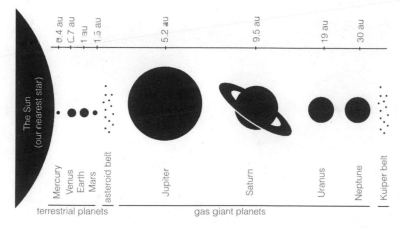

Figure 1 Our Solar System. 'Astronomical units' (au) are used to compare the immense distances between the planets. 1au is the distance from the Earth to the Sun. (Due to the huge size difference between the terrestrial and gas giants, this image is not to scale.)

In the early 1990s, we knew of eight planets:

> Mercury,
> Venus,
> Earth,
> Mars,
> Jupiter,
> Saturn,
> Uranus,
> Neptune.

And the dwarf planets, Ceres (within the asteroid belt) and Pluto (within the Kuiper belt).

The first four are the terrestrial planets, with rocky surfaces and thin atmospheres. The next four are the gas giants, 15–300 times more massive and engulfed in atmospheres thousands of kilometres thick.

But these were not the only worlds out there.

The Blind Planet Hunters

It was six men of Indostan
To learning much inclined,
Who went to see the Elephant
(Though all of them were blind),
That each by observation
Might satisfy his mind.

The Blind Men and the Elephant by John Godfrey Saxe,
based on the Indian fable

There is an Indian fable about six blind men examining an elephant. Each reaches out to touch a different part of the unknown animal. One man finds the smooth membrane of the elephant's ear. Another holds the curved tusk, while a third grasps at the thin tail. The fourth man touches the elephant's trunk and the fifth wraps his hands around one leg. The final man presses his palms against the elephant's broad flank. An argument then erupts over what an elephant truly looks like, since each man has only found part of the truth.

'What would make you throw my book out of the window?'

Winter sunlight was streaming through one such possible glass pane on the third floor of the University of Washington's Department of Physics. Outside, the damp Seattle skyline was creating an impressive panorama, but all I could think about was a crushed copy of my book lying in a puddle.

In the chair opposite me was Tom Quinn, a thickly bearded astrophysicist who had spent decades working on models for planet formation. I had talked his ear off for the last 10 minutes listing all the world-changing feats I wanted my magnum opus to accomplish. Now we came to the crunch point: what would make a renowned expert in planetary science dismiss a book on alien worlds as rubbish? I expected Quinn to respond by ticking off essential topics on his fingers. Number one on his list would surely be the importance of hot Jupiters; the first discovered planets around stars like our own that smashed all previous formation theories into dust. Next in line might be the mysterious super Earths, whose sizes do not match anything circling our Sun. Are these miniature gas planets with suffocating atmospheres, or rocky worlds in size XXL?

Perhaps Quinn would mention the planets that orbit two twinned stars like the fictional home world of Luke Skywalker, or the other extreme where the planet has no sun at all. Then there were the worlds whose paths around their stars was so elongated that their seasons swing between fireball and snowball, worlds where the sun never sets, worlds entirely under water, or with a ground of molten lava. Alternatively, Quinn could say that the next big discoveries would be planets like our own Earth, with rugged coastlines that harboured strange forms of life.

Quinn did not make a list. Instead, he spoke frankly.

'Our knowledge of planet formation is not complete,' he said. 'We have seen only a tiny fraction of what is out there. If you presented what we know as a complete picture of what exists, then I would throw your book out of the window.'

Quinn's point was that the secrets of planets shared the conundrum of the blind men and the elephant. The myriad of worlds in our cosmos is an unseen creature that we are struggling to understand through the small sections we have so far uncovered.

The star speed gun

In 1968, Michel Mayor fell down an icy crevice and almost missed discovering the first planet orbiting another sun.

Mayor was an explorer. Born in 1942 in Lausanne on the shores of Lake Geneva in Switzerland, he grew up in a family

with a love of outdoor activities. This expanded into a rather
risky passion for high-altitude skiing and climbing, which led
to him hanging precariously off an ice-laden ledge 26 years
later. Perhaps it was this love of high places that led to Mayor's
obsession with the motions of the stars.

For his doctoral thesis at the University of Geneva, Mayor
was searching for small deviations in a star's course due to the
pull of gravity from our Galaxy's spiral arms. It was a study
that required recording the velocity of stars to impressive
levels of precision. As Mayor worked on techniques to
improve these measurements, the changes in the star motions
that could be detected became steadily smaller. Eventually,
even tiny wobbles of the star could be seen; wobbles caused
by an object vastly smaller than the star itself – a nudge from
an unseen planet.

The problem with detecting planets is that stars are big and
bright. Even the most massive planet in our Solar System,
Jupiter, reflects just one billionth of the Sun's light. This makes
planets immensely hard to spot when they orbit another star
whose own light is only a pinprick in the sky. Yet the technique
Mayor was working with did not require astronomers to see
the planet directly. Instead, they would measure the star
wobbling as the planet orbited around it.

When talking about orbits, we normally think about a
smaller object circling a stationary more massive body, for
example the Earth circling the Sun, or the Moon circling the
Earth. In actual fact, both bodies pull on one another and
therefore both move. The pair orbit around their *centre of mass*;
a balance point in space between their two gravitational pulls.

This is similar to sticking two erasers on either end of a
pencil and trying to balance it across your finger. If both
erasers weigh the same, then the balance point is at the pencil's
centre. This is the case when two equal-mass stars form a
binary star system. The stellar twins circle a point halfway
between their positions. Alternatively, if the erasers have
different masses, the balance point moves along the pencil to
the end closer to the heavier eraser. Pluto's giant moon,
Charon, has almost 12 per cent of the dwarf planet's mass.

Their centre of mass lies roughly 1,000km (620mi) above Pluto's surface, and just under 17,000km (10,500mi) from the surface of Charon. This causes Charon to make a large circle and Pluto to make a smaller circle as they both orbit this balance point.[*] As our own Moon has only 1 per cent of the mass of the Earth, their centre of mass is approximately 1,700km (1,050mi) below the Earth's surface. The Earth still orbits this position, but its motion is more of a jiggle as it moves around a point within its interior.

When we reach a star and a planet, the mass difference is so huge that the centre of mass lies very close to the star's physical centre. The planet then moves in a large circle almost directly around the star, while the star's orbit consists of a tiny wobble.

At the end of 1994, Mayor's graduate student, Didier Queloz, was working alone at the telescope when he saw such a wobble. This small movement was for a star 51 light years away in the constellation of the Winged Horse, Pegasus. It was the signal of an extrasolar planet, or *exoplanet*: a planet beyond our Solar System.

The mechanics behind detecting such a tiny wobble are similar to listening to an ambulance siren. As the ambulance approaches, the distance between you and the siren is reduced. This compresses the sound waves, which shortens their wavelength, causing a rise in the siren's pitch. When the ambulance moves away, the sound waves stretch out and the pitch drops. This is known as the *Doppler Effect*.

Exactly the same phenomenon happens with the star's light. As the star moves slightly towards the Earth during its orbit with the planet, its light waves compress and shorten in wavelength, making the light bluer. When the star moves back away from the Earth, the light waves stretch and become redder. As the planet and star orbit their centre of mass, the star's light oscillates between bluer and redder wavelengths in keeping with the star's slight wobbling motion.

---

[*] During its approach in 2015, NASA's *New Horizons* space probe imaged Pluto and Charon during their orbit. The impressive animation can be found on the mission's web pages.

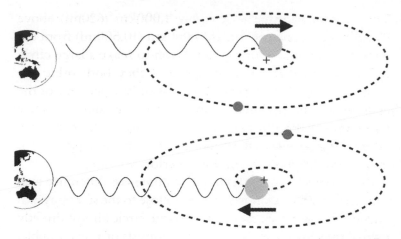

Figure 2 Finding planets using the radial velocity technique. The mutual pull of the planet and star on one another cause both to orbit about their centre-of-mass (+). When the star moves away from the Earth (top image) its lightwaves are stretched and become redder. As the star moves back towards the Earth, the lightwaves compress and become bluer. This change in the star's velocity reveals a hidden planet.

Another way of looking at the same problem is to consider particles of light. The star is like a person throwing light balls to you at a steady rate. As the star moves towards you, the distance between you and the star decreases and you receive the balls more rapidly. This is the decrease in wavelength that makes the light more blue and raises the siren's pitch. When the star now moves away from you, the distance increases so the balls take longer to arrive. The wavelength has now stretched out and reddened.

Measuring this shift in the wavelength maps the change in the star's motion as it wobbles towards and away from the Earth. The motion in our direction is known as the star's 'radial velocity' and gives the planet-finding method the name the *radial velocity technique*, or sometimes the *Doppler wobble*.

From the time it took for the star to wobble back and forth, Mayor and Queloz could measure the duration of the planet's orbit. From this, they could deduce how far the planet was from the star. Meanwhile, the magnitude of the star's wobble hinted at the mass of the planet; the more the star had to move in its orbit, the further away the centre of mass balance point had to sit and thus, the weightier the planet.

In fact, the planet mass estimated by the radial velocity technique is always the minimum possible value. This is because only the motion directly towards or away from us causes the light wavelengths to compress or stretch. Any part of the star's wobble not directed our way will go undetected.

This is similar to tracking the motion of a hot-air balloon by following its shadow. The moving shadow shows the balloon's motion parallel to the ground, but does not reveal if the balloon has climbed or decreased in height. If you were using the shadow's motion to estimate how much fuel the hot-air balloon had burned, your guess would often be too low since fuel used to increase the balloon's altitude was not included. Likewise, if the planet and star orbit at an inclined angle to the Earth, only part of the star's wobble will be in our direction and detectable. This causes the force from the planet to be underestimated, giving a mass lower than the true value.

Mayor and Queloz were observing at the Observatoire de Haute-Provence telescope in the south of France. By the end of 1994, the two scientists had 12 measurements for the radial velocity of the star 51 Pegasi and realised they had something big. But then they hesitated. Previous attempts to find something as tiny as a planet had given planet searches a terrible name. The past 50 years had been littered with mistaken reports that had to be retracted on closer examination. Was this definitely a planet, or were they just seeing a small variation in the atmosphere of the star as it rotated?

There was another problem. When they calculated the minimum mass and the time for one orbit of the planet, the numbers made no sense at all.

The supposed planet was a world at least half the size of Jupiter, making it around 150 times the size of the Earth. Its huge bulk meant that this was a gas giant planet like our big four – Jupiter, Saturn, Uranus and Neptune. These are worlds that probably harbour solid cores, but are dominated by massive atmospheres thousands of kilometres thick. All of our gas planets are in the outer parts of our Solar System. It was a location planet-formation models declared universal; to form a gas giant a large amount of material was needed. This simply does not exist close to the star, where the heat evaporates away a large chunk of the possible solids. Gas giants must therefore always be far away from the star. Yet, this new planet was not far from the star. In fact, it was much closer to 51 Pegasi than Mercury is to our Sun. One year on this new world lasted just four days. Surely, this had to be a mistake?

Mayor and Queloz waited and performed more observations of 51 Pegasi. In July 1995 they added eight new measurements. Looking at their data they finally became sure. Despite all the odds, this was a real planet.

On 6 October that year, Mayor attended a workshop in Florence, Italy. He had registered late for the conference, so was invited to join the round-table discussion, where he would be able to give a brief five-minute presentation. Before the meeting began, rumours became to circulate about what Mayor had to present. The organisers raised his talk time to 45 minutes.

When Mayor stood up, he announced the first extrasolar planet to be discovered by the wobbling motion of a sun-like star. And with that, the floodgates opened to finding dozens of new worlds.

🪐

Mayor's boiling alien world became 51 Pegasi b, a moniker that comes from the star name, 51 Pegasi, followed by a lower-case letter. Convention leaves the small-case letter 'a' for the star itself, giving the first planet discovered in a planetary system the letter 'b'. Subsequent planetary siblings then become 'c', 'd', 'e' and onwards. Should the star belong to a binary system with two orbiting stellar siblings, then capital 'A' and 'B' are tacked on to refer to each of the stars.

The star name can have a variety of often unwieldy sources. 51 Pegasi is the 51st star assigned within the Pegasus constellation. Other stars are named after the astronomical catalogue that lists them. For example, Gliese 1214 is the 1,214th star in the *Gliese* catalogue, while BD+20594 is from the *Bonner Durchmusterung* catalogue. As we will see later, many stars with orbiting planets are named after the instrument or survey that detected the new world.

While there had never been real doubts that planets were common around other stars, 51 Pegasi b marked the beginning of being able to reliably find these alien worlds. Then, in 1999, another detection occurred that would herald the start of these new planets being found in droves.

The silhouette of Venus

The home of the Astrophysics Department at the University of Oxford has the dubious distinction of being one of the ugliest buildings in the historic city. Yet the crowd that gathered on the rooftop on 8 June 2004 was ignoring the concrete architecture. Instead, it was intently focused on a makeshift screen on to which a pinhole camera was projecting a shadowy image of the Sun. As the clock ticked past noon, a darker silhouette began to creep across the fuzzy surface. It was the planet Venus, transiting across the Sun for the first time since 1882.

A transit occurs when a celestial body passes directly between a larger body and the Earth (or another vantage point), hiding a small part of it. The most extreme example of such an event is a total solar eclipse, in which the Moon briefly obscures all the Sun's light. While the diameter of Venus is almost three-and-a-half times larger than that of the Moon, its more distant location results in it only blocking out about 0.1 per cent of the Sun's light. With such a tiny dip, Venus's transit would pass unnoticed to anyone not armed and ready with observational equipment. This did not happen until 1636.

Johannes Kepler was a German astronomer who died six years before the transit of Venus he failed to predict. He is best known for recognising that planetary orbits are elliptical, rather than circular, and deriving three laws that describe their motion. These painstakingly accurate observations of how the Solar System's planets moved also led to the first estimates of when Venus would cross the Sun's face.

The required alignment of the Sun, Venus and the Earth makes such transits rare. Venus photobombs the Sun in pairs of events separated by over a century of time. Kepler's calculations estimated that Venus would just miss crossing the Sun's surface in 1836. These figures were updated by British astronomer Jeremiah Horrocks, who not only realised that Venus would transit, but proceeded to make the first recorded observations of the event along with his friend, William Crabtree. Ironically, Horrocks's use of a telescope to focus the Sun's image meant that his equipment was more sophisticated than the set-up on the roof of the Oxford Astrophysics Department building 168 years later.

While Venus only transits our Sun on rare occasions, across the night sky countless planetary transits are occurring. Yet finding these would mean detecting the wink of a planet across a pinprick of starlight.

Exoplanet history – Australian planetary scientist Stephen Kane would later tell me over a beer – is split into two parts: before the discovery of HD 209458b and what would come after it.

HD 209458b is another Jupiter-sized world that sits close to its star, completing its orbit in a swift three-and-a-half days. Its ungainly name is an example of a particular astronomical listing: 'HD' is for the *Henry Draper* catalogue and '209458' for the star's coordinates in the sky. Like 51 Pegasi b, this position places HD 209458 in the constellation of Pegasus but three times further away, at a distance of 150 light years from us.

The planet had first been found via its star's wobble using the radial velocity technique. However, such a big planet

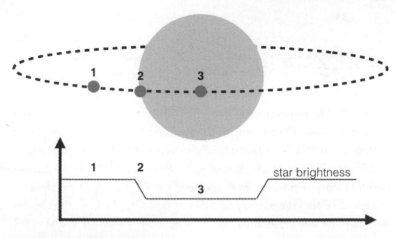

Figure 3 Finding planets using the transit technique. As the planet crosses the star's surface, it blocks part of the light and the star's brightness drops.

close to its star stood a tantalising chance of making a detectable transit. Excited by this possibility, two teams began to monitor the light from HD 209458.

For stars more distant than our Sun, the defined outline of the silhouette of a planet's shape as it transits cannot be seen. Instead, the star's light shows a small wink-like dip. Such a dimming is tiny. Even for a Jupiter-sized gas giant such as HD 209458, the dip in the star's light is only of the order of 1–2 per cent. For a planet the size of the Earth, this drops to below 1/100th of that 1 per cent.

Despite these challenges, both teams focused on HD 209458 spotted a telltale reduction in the star's brightness that lasted a few hours. Their results were both published in the same edition of *The Astrophysical Journal* in December 1999. The dips observed in the starlight corresponded exactly with the periodic variations in the star's position found via the radial velocity technique: the first transiting exoplanet had been found.

This new planet-finding method became known as the *transit technique* as it sought the transiting of a planet across a star's surface. Rather than the planet mass estimate yielded by

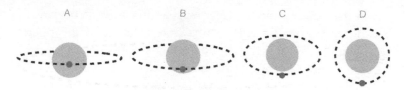

Figure 4 The importance of the orientation angle of the planet's orbit around its star. Planets on orbits C and D do not cross the star's surface when seen from Earth, and can never be found with the transit technique. Planets in A, B and C can potentially be found by the radial velocity technique. In the edge-on orbit A, the measured planet mass will be the true mass of the planet. In orbits B and C, the planet mass will be underestimated because only part of the planet and star's motion is towards the Earth. If the planet can be found with both the transit and radial velocity techniques in orbits A and B, the planet mass and radius can be measured to give an average density. The planet in orbit D cannot be found by either technique and relies on another method, such as direct imaging.

the radial velocity technique, the transit technique reveals the planet's radius. A bigger dip in the starlight corresponds to a larger planet. This made HD 209458b the first exoplanet whose size was known.

As well as planet size, this technique also provides the orbit's orientation. By knowing the time it takes for a planet to cross the star (the duration of the light dip), and the time to orbit the star (the time between light dips), the planet's path can be mapped out. This removes the uncertainty in the radial velocity mass measurement, so the combined techniques can reveal an accurate mass and radius for the new world.

Measurements of both the mass and radius provide more than just the physical extent of the planet. Together, they give an estimate of the planet's average density; a value that is a clue to the nature of the planet itself.

A rocky, terrestrial planet such as the Earth has a high density of $5.51 \text{g}/\text{cm}^3$. However, the Earth's iron core is substantially denser than this value, while its surface material is less dense, so the value quoted for the whole planet is an average over its composition.

For the giant, Jupiter, its sizeable mass is accompanied by an even more sizeable radius due to most of the planet consisting of hydrogen gas. This gives the planet a very low average density of $1.33g/cm^3$.

In the case of HD 209458b, these properties for the planet continued to pile on surprises to its already unusually hot orbit. The planet was two-thirds as massive as Jupiter but over a third larger. This yielded a density of just $0.37g/cm^3$. It is certainly a Jupiter-like gas giant, but a bloated one.

Detecting both the radial velocity wobble and transit light dip is far from easy. Not all planets transit their star's surface and others do not create a wobble that is big enough to be differentiated from the star's own lively variations. However, having methods to probe the average structure of exoplanets was a huge boon – a boon big enough to encourage a much bigger exoplanet project.

Late at night on 7 March 2009, a carrier rocket blasted away from the launch pad at the Cape Canaveral Air Force Station in Florida, US. It carried on board the first space telescope dedicated to planet hunting.

The telescope was named after Johannes Kepler, the astronomer who had painstakingly calculated the motions of the planets in our solar system. His contribution to predicting the transits of our closest planets made him the eponym for a telescope that would see the transits of thousands.

Once in space, the Kepler telescope manoeuvred into position to trail the Earth around the Sun. Then, on 7 April, the protective dust cover was ejected and Kepler saw its first light. The 1.4m (55in) mirror trained on a patch of our Galaxy in the constellations of Cygnus and Lyra; an area rich in stars allowing Kepler to observe more than 100,000 pinpricks of light at a time.

By spotting the dips in the stars' light, the Kepler Space Telescope used the transit technique to find exoplanets crossing in front of their star. Freed from the scattering effects of the Earth's atmosphere, Kepler was far more sensitive to the tiny dimming of starlight than ground-based observations.

It would prove to be a wild success. At the winter meeting of the American Astronomical Society in January 2015, the Kepler team announced their 1,000th confirmed planet discovery. This was in addition to more than 4,000 *planet candidates* whose existence was suspected, but for which further observations were required in order for scientists to be sure. The official objective of the mission had been to search out Earth-like planets, but Kepler's real gift has been uncovering the diversity and sheer number of worlds in our galactic neighbourhood. Within 20 years, we have moved from basing all planet-formation theories on our single Solar System, to comparing it against more than 500 different planetary systems.

Both the transit and radial velocity techniques are most sensitive to big planets orbiting close to their stars. These planets block out the most light, are the most likely to transit across a star's surface, and are bulky enough to give the star an appreciable wiggle. The result is that we know far more about worlds on short orbits than the ones on the outskirts of their systems.

These two techniques are not the only ways of finding extrasolar planets, but they are the most prolific. As I write this, there are 3,439 confirmed extrasolar planets, and 3,314 have been found through at least one of those two methods.[*]

This book is the story of those 3,439 planets. It is the travel log of how they came to form from dust particles to worlds so diverse that even Hollywood has failed to be weirder. At least one of those worlds developed a sentient life form capable of asking how that happened. That life form needs to remember one thing: everything in this book should be questioned.

We're not done.

---

[*] I can guarantee that number will now have changed. The current count can be found on NASA Exoplanet Archive: exoplanetarchive. ipac.caltech.edu.

PART ONE
THE FACTORY FLOOR

CHAPTER ONE

The Factory Floor

An hour past midnight on 8 February 1969, a fireball lit the skies above the northern Mexican state of Chihuahua.

'The light was so brilliant that we could see an ant walking on the floor,' Chihuahua newspaper editor Guillermo Asunsolo told the USA's *Washington Post*. 'It was so bright we had to shield our eyes.'

The burning rock hurtled through the atmosphere before exploding over the village of Pueblito de Allende, scattering debris over a 250km^2 (96mi^2) region. It was a sight to inspire apocalyptic fears of world destruction. Yet, far from being an omen of our death, this burning spectacle was a memento from our birth.

Rocks that enter the Earth's atmosphere from space are known as meteors. Hitting the Earth's atmosphere is a dangerous pastime for a rock, since the air provides a much higher resistance to its motion than the vacuum of space. As the meteor slams into the atmosphere, the air is rapidly compressed, which causes the temperature to soar. This heated air around the incoming meteor lights up, turning small sand grains into shooting stars, while the rare large boulders become fireballs. Such extreme cooking produces a high chance of complete incineration, and most meteors never make it to the Earth's surface. The ones that do survive this perilous journey have their bravado recognised by being reclassified as *meteorites*.

The spectacular entry of the Allende meteorite (named after the village over which it exploded) earned it instant fame. Scientists descended on the debris field, assisted in their search for meteorite pieces by local residents and schoolchildren. Among the researchers was a field team from the Smithsonian Institution in Washington DC, which gathered approximately 150kg (330lb) of meteorite material and distributed it to 37 laboratories in 13 different countries within a few weeks of the fall. In total, more than 2 tonnes of material was collected, ranging from 1g fragments to an enormous chunk weighing 110kg (240lb). This massive quantity suggested that the meteor was the size of a car before its explosion. The successive collection and distribution of the rock would help the Allende meteorite earn the title of the 'best-studied meteorite in history'. However, its size was not the only anomaly that made it important.

In early 1969, research laboratories throughout America were on tenterhooks. They were waiting for lunar rocks from *Apollo 11*'s historical Moon landing, when another example of a rock from space slammed into their neighbour's backyard. With the laboratory equipment already primed for extra-terrestrial material analysis, pieces of the Allende meteorite recovered from the fall site were examined to reveal that this was no common space rock. Rather, the meteorite's grey-and-white dotted composition was that of a *carbonaceous chondrite*: a class of meteorite that accounts for less than 5 per cent of all meteorite falls. It is a class that consists of the very first objects to form in the Solar System, and the Allende meteorite remains the largest example ever to have been found on Earth.

Its early origins make the discovery of a carbonaceous chondrite akin to holding a baby picture of our most distant ancestor. The rock was formed at the very start of our planet's story, but unlike the Earth it failed to gain sufficient mass to grow into a world of its own. From this physical snapshot of our own beginning, we can precisely date the birth of our planetary neighbourhood precisely.

Laboratory analysis of meteorites reveals that they contain elements that are radioactive, with atoms that can sponta-neously change into those of a different element. Due to the random nature of this radioactive decay, it is not possible to say exactly when a particular atom will change. However, for a large number of atoms, scientists can measure with some certainty the time it takes for half of them to decay. This time period is known as the *half-life* of the element. What this means is that if we know what fraction of the radioactive element has decayed, we have a clock with which to calculate how much time has passed.

One such radioactive element commonly found in meteorites is rubidium–87 (written as ^{87}Rb). The '87' refers to the mass of the rubidium atom nucleus; the central region containing positively charged particles named 'protons' and particles called 'neutrons', which have the same mass as a proton, but with no electric charge. When an atom of ^{87}Rb decays, one of its neutrons becomes a proton in a process known as beta decay. The result is an atom of strontium–87 (^{87}Sr), which has a nucleus with the same mass as ^{87}Rb, but with one extra proton and one less neutron.

The time taken for half the atoms of ^{87}Rb to decay into ^{87}Sr is 48.8 billion years. This is a good duration for measuring planet-formation timescales. If the half-life was very short (say a few years), then the ^{87}Rb atoms would be long gone before the rock being examined had made it to Earth. On the other hand, if the half-life was very much longer, there might not have been enough ^{87}Sr atoms to measure. Roughly speaking, the time period to be measured needs to be between a tenth of a half-life to 10 times the half-life for the radioactive dating to be accurate.

By measuring the current quantity of ^{87}Rb atoms in the meteorite and the number of ^{87}Sr atoms that the rubidium decay must have produced, scientists can calculate what fraction of atoms must have decayed since the meteorite's birth. Combined with the ^{87}Rb's known half-life, this gives the time that has passed since the rock's first formation.

For a carbonaceous chondrite such as the Allende meteorite, this age marks the very starting point of our planet's history: a number that comes out as 4,560,000,000 years.

The planet-forming disc

While the Allende meteorite tells us the date of our planet's birth, exactly what existed at that time is a bigger mystery.

Rather than being a clear ancestral family photograph, a carbonaceous chondrite is more like a close-up selfie of a distant cousin, with a date scrawled in the lower corner. Without a better image of the conditions that started our planet's formation, it is impossible to estimate the likelihood of us ever finding a second similar world.

Although the lack of a better family photographer is disappointing, we do know one fact about our birth epoch: 4.56 billion years ago, our own Sun had just been born. As it happens, this single connection with recent star formation is the only information required to reveal what is needed to build a planet.

Stepping back another few million years before the primitive meteorite was formed, we end up in one of the coldest places in the Galaxy. This was the Sun's own nursery: a cloud of gas so cold that its temperature measured a staggeringly low -263°C (-440 °F). Such cloud nurseries are the birth places for all the Galaxy's stars. Made predominately of hydrogen, these cradles have masses of around 1,000–1,000,000 times that of the Sun. Their creation within the constantly moving Galaxy ensures that their gas is not distributed smoothly throughout the cloud, but shuffles around like a lumpy mattress, gathering in dense pockets of gas known as *cores*. The high mass concentrated into a small space causes gravity to force the core to contract, increasing its density still more and speeding the collapse. As the gas falls in on itself, it heats and an embryonic star – a protostar – is born.

While gravity may be congratulating itself on compressing material into a star, it is not the only force at work. Knocked about by the Galaxy's own rotation and interactions with neighbouring clouds, gas within the nursery cloud rotates. Just as riding a children's roundabout causes you to be thrown outwards, so this spin supports the gas against gravity's pull. The extra force holds the fastest rotating gas in the core away from the collapsing protostar. The result is like spinning dough into a pizza, and the star becomes surrounded by a rotating disc of gas.

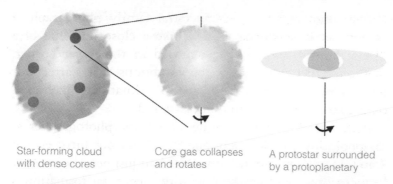

Star-forming cloud
with dense cores

Core gas collapses
and rotates

A protostar surrounded
by a protoplanetary

Figure 5 The formation of the planet-forming protoplanetary disc. Stars are born in dense cores in cold clouds of gas. These cores rotate as they collapse under gravity, producing a young protostar surrounded by a disc of gas and dust.

As the gas settles and begins to cool, dust particles condense within the disc, like ice crystals solidifying out of water vapour. These tiny grains join a splattering of dust already in the gas cloud to become the first solids around our Sun. These are the earliest beginnings of planet formation. From these minuscule building blocks, steadily larger objects can be assembled in the rotating gas and dust factory, which is now known as the *protoplanetary disc.*

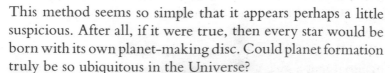

This method seems so simple that it appears perhaps a little suspicious. After all, if it were true, then every star would be born with its own planet-making disc. Could planet formation truly be so ubiquitous in the Universe?

One simple test for this is to ask whether protoplanetary discs can be seen surrounding young stars today. The problem with this is that the disc does not shine. Unlike the central star, which is busily warming up to become a ball of fiery glory, the surrounding dusty disc cannot produce its own light. However, the dust will absorb the energy coming from the star. In the same way that a car's bonnet excels at absorbing the Sun's rays to get scorchingly hot on a summer's day, energy from the star's light will heat the disc dust. The warm dust then releases its heat as low-energy infrared radiation.

The human eye is not sensitive to infrared, but cameras that can detect it are easy to find. Unfortunately, the perfect piece of equipment to image the heat from a night-time robber cannot be turned on the skies to spot a protoplanetary disc. This is because while the disc is heated by its central star, its temperature can still drop to far below anything found on Earth. In order for the camera's own heat to not interfere with the detection, the equipment must be cooled to temperatures colder than even a stellar nursery. Additionally, the Earth's own atmosphere is better at absorbing infrared radiation than the aforementioned robber was at running off with your new TV set. Therefore, the best place to put such an instrument is in space.

While substantially easier to keep cold, space telescopes hunting in the infrared still require cooling. This is typically done with liquid helium, which slowly evaporates by absorbing away the surrounding heat to keep the telescope at -270°C (-454°F). Once the helium has evaporated entirely, the telescope warms slightly to a relatively balmy -244°C (-407°F).

Two such telescopes with missions to hunt for discs around young stars were the Infrared Space Observatory and the Spitzer Space Telescope. The first of these was launched in 1995 by the European Space Agency. It continued operating until 1998, at which point the helium coolant ran out. The Spitzer Space Telescope is one of NASA's four Great Observatories, a prestigious group that also includes the Hubble Space Telescope. It was launched in 2003, and its coolant finished in May 2009, but it continues to operate at a reduced capacity with the warmer temperature. The results from these instruments have been unequivocal: stars younger than a million years old are all surrounded by dusty discs. If the planets form from such a pile of parts, then every new star does indeed have the ability to build new worlds.

Yet these investigations also uncovered another result. While the youngest stars have discs, only 1 per cent of stars beyond the age of 10 million years retain these planet-making kits. This leads to a single conclusion: there is a clock ticking for planet formation.

There are several destructive methods for taking out a protoplanetary disc. The most exciting prospect is that the entire disc will be turned into planets, producing a miasma of new worlds. Unfortunately, observations of both our own Solar System and known exoplanet systems show that their total eventual planetary mass is only 1 per cent of the initial disc mass, leaving a question mark about the remaining 99 per cent.

Another possibility is that nearby stars pull on the disc with their own gravity, stripping it away from its host sun. This process may occur occasionally, but it isn't thought to be a common enough event to be responsible for the complete obliteration of all protoplanetary discs; stars are typically just too far apart. The destruction must therefore be an inside job: that is, the forming star and disc system destroys itself.

Some of the blame for the destruction is due to friction within the disc. We can picture this by imagining the disc as a consecutive series of running tracks around the star. Gas in the inner track pulls ahead of gas in the adjacent outer track. Friction between the tracks slows the speedy inner gas, reducing its rotational support against the protostar's gravitational pull. The outer gas increases in speed from the forward tug of the inner gas track, but is slowed in turn by its own neighbouring outer track. As the rotational support diminishes in the disc, the gas and suspended dust fall towards the star.

This inward flow is called accretion and it is certainly responsible for removing a portion of the disc. It does not represent the whole answer, though, as the process is rather slow. To remove the outer parts of the disc via accretion would take several thousand million years, yet observations tell us that we only have about 10 million years to get the job done. More damningly, very few discs are seen in the process of partial destruction, suggesting that the actual demolition time is 10 times shorter still, and must take place almost simultaneously across the whole disc. This last part is a particular problem since accretion turns out to be fastest closest to the star, so eats away at the disc from inside out. What is needed is a second, faster destructive force, and this is provided by the star itself.

Like a teenager going through a painful adolescence, the progression from the young protostar to a fully fledged sun is an angry process. In the case of an intermediate-mass star like the Sun, this rebellious stage is known as the *T-Tauri phase*, named after the first star to be observed at this awkward moment, in the constellation of Taurus, the Bull. In a similar manner to throwing insults at protective parents, T-Tauri stars throw out damaging radiation in the form of high-energy ultraviolet and X-rays, along with winds packed with a blowtorch of high-energy particles. These hit the upper gaseous layers of the disc and heat it. Close to the star, the result of this energy bombardment is just a very hot disc. Further out, however, the pull from the star's gravity is weaker, and this energy can be enough to allow the disc gas and smallest dust grains to escape as a wind. This is known as *photoevaporation* (literally 'evaporation from photons', the particles of radiation), and is the process thought to be responsible for the destruction of the main part of the disc. Close to the star where the gravity is sufficient to hold against photoevaporation, accretion then finishes the job.

Once the gas disc has been removed, the only freewheeling pieces will consist of planets and other solid objects big enough to have avoided being carried off with the gas. Most gas still in the system is now already part of a planet, where it can be held in place by the planet's gravity. Since our Solar System contains four planets with a huge part of their volume in their gaseous atmospheres, we know that the planetary neighbourhood must be nearly finished by the time the disc is destroyed. This gives us approximately 10 million years to go from a pile of dust particles 10 times smaller than a grain of sand, to an entire world that looks as though it might one day host life.

At this point, it would not be unreasonable to suggest that this seems to be a near-impossible task. So much so, that one might propose that the discs visible around young stars are not planet-forming entities at all, but merely the dusty placentas of new-born stars. One way to test this hypothesis is to ask how much material must have been in the Sun's protoplanetary disc to produce the Solar System. If this is vastly different from the mass of discs observed around young stars, then this progression from dust to planets must surely be bunkum.

Had we mimicked this assembly process by building a model of the Solar System out of LEGO bricks, the job of finding the quantity of starting materials would be easy. By breaking up the construction and counting the number of plastic bricks that had been used to form the planets, we could accurately state the number needed to complete such a project. However, when doing this for the protoplanetary disc, we have the problem that a compulsive kleptomaniac – the Sun – always steals a large fraction of the bricks during the assembly process.

If all the planets in the Solar System were broken down and smeared out to form a disc, the resulting system would be rich in iron and silicate compounds containing silicon, magnesium, carbon and oxygen, with ices being abundant further from the Sun. These were the heavier elements that condensed most easily out of the gas and into solids, forming dust and then (our mechanism proposes) larger rocks and planets. While lighter elements such as hydrogen could bond on to the dust grains to form solid compounds such as ice or be captured in planetary atmospheres, most were evaporated from the disc by the young Sun's radiation.

Reporting this deficiency of light materials to an insurance company would probably result in being told that our story was unbelievable unless we could prove the quantity we originally possessed. This would seem a tall order, except that if the disc formed from the same stellar-nursery gas as the

Sun, we have a comparison point for the material it must have initially contained: the Sun itself.

If the toy Solar System model were made from a box of coloured bricks and the brick stealer had a strong preference for red, assessing the number of bricks used in the construction would be much easier. By knowing that the bricks used in the assembly came in an equal mix of red, green and blue colours, we could estimate the number of missing red bricks from the total numbers of the other two colours. For example, if we disassembled our model to find it contained 100 green bricks, 100 blue bricks and five red ones, then it would be reasonable to assume that 95 red bricks had been taken, and our original number totalled 300 bricks.

It is this same technique that can be used to find the quantity of missing elements in the protoplanetary disc. Forming from the same gas core, the disc and Sun must originally have had the same ratio of elements. Like the red bricks, the volatile elements in the disc have been removed, but their number compared to the heavier elements must be the same as in the Sun. To estimate the original disc mass we can therefore add lighter elements to the disc of crumbled planet parts until the ratios between each element equal the solar amount. This does assume that the Solar System was perfectly efficient in gathering into planets the heavier elements we do see into planets; in reality, some of this mass will have been lost during the adolescent Sun's T-Tauri temper tantrums. Nevertheless, it gives us an absolute minimum for the mass needed to form the Solar System. That value is known as the *Minimum Mass Solar Nebula* (MMSN), and it turns out to be around 3 per cent of the Sun's mass. This also happens to be similar to the estimated mass of the observed discs around young stars.

Particularly spectacular evidence that the protoplanetary disc can yield a solar system full of planets came from another rock. On 9 May 2003, the Japan Aerospace Exploration Agency launched an unmanned spacecraft to land on an asteroid named Itokawa.

Asteroids are rocks typically a few kilometres to a few hundred kilometres in size that mainly sit between Mars and

Jupiter. Collisions between asteroids can send fragments scattering towards the Earth, some of which end up smacking our planet as meteorites. One such collision early in Itokawa's history had pushed the asteroid on to a new orbit closer to the Earth, making it an easy-to-reach target for a spacecraft.

Japan's spacecraft was called *Hayabusa*, translating to *peregrine falcon* in English, and it both photographed the 550m (800ft) long Itokawa and brought home samples from its surface in June 2010. Images from the mission showed a peanut-shaped object that consisted of a rubble pile of many different-sized pieces. Rocky boulders and dusty granules were held together loosely by Itokawa's gravity, which was not strong enough to pull the asteroid into a dense round ball. Missions to different asteroids supported this view of a collection of irregular lumpy rocks. Such a morphology must have developed from the collision and sticking of the smaller visible pieces, evidence of this factory assembly mechanism in action. The result is the planets and the remaining asteroid rubble; the dust on the factory floor.

This dusty gas disc is therefore truly the planet-building factory floor. From here we begin an assembly that will take us from grains of dust to eight new worlds between 10,000 billion and 100,000 billion times greater in size. It is the greatest construction process in the Universe, and it has taken place around each and every star you see in the night sky.

CHAPTER TWO

The Record-breaking Building Project

In August 2013, the US city of Wilmington in Delaware saw the birth of a record-breaking monument: a 34.44m (113ft) tower made entirely from interlocking plastic LEGO bricks. The brightly coloured creation was assembled by pupils from 32 schools in the Red Clay district, each of whom pieced together separate sections of the tower. These larger segments were then combined by a construction crew and crane in a grand finale that was officially recognised by the *Guinness Book of Records*. In total, almost 500,000 LEGO bricks were used, and the end product beat the previous world-record holder by nearly 2m (6½ft).

It was a project that demonstrated something the Universe has known for billions of years: in order to build something truly big, you start small and work upwards. In the case of our own Solar System, this involves taking the microscopic dust grains around a young star and sticking them together to form a planet.

Despite the fact that we know this assembly process must work, planetary scientists were faced with two baffling problems. First, it was not remotely obvious how dust particles would stick together. The Itokawa asteroid kept its rubble pile of rocks in place by its own gravity. But the strength of gravity depends on the object's mass. A rocky body smaller than roughly a kilometre in diameter would not have enough bulk to produce a decent sticking force. The result would be like pressing dry sand together on a beach; the grains slide apart as soon as you release your grip.

Second, it was unclear how any sticking mechanism could work fast enough to build our Solar System before the Sun destroyed the protoplanetary gas disc. The observations of protoplanetary discs around young stars gave them an upper time limit of 10 million years. Within this span, an assembly process would have to take dust grains a tenth of the size of sand, and create a young planet with enough mass to hold on to a gas atmosphere as the rest of the disc evaporated.

In short, this process was akin to being given a box of bricks and told to build a tower, only to find that the bricks were completely smooth and the box would be confiscated straight after lunch.

●

While even a record-breaking tower of building blocks can be comfortably measured in metres, the Universe's construction projects happen on a rather larger scale. To avoid having to deal with numbers of a ridiculous size, let's take a small detour to find a more practical unit of distance to explore the Solar System.

It is perfectly possible to discuss the positions of the planets in terms of metres or kilometres, but it is hard to make sense of numbers once they reach an obscenely large number of digits. For example, the Earth is 149,600,000km (92,960,000mi) from the Sun, while Jupiter sits out at 778,340,000km (483,640,000mi). Compared with your average run to the supermarket, both these distances come under the category of *disturbingly big*, and it is difficult to quickly tell how much deeper into our Solar System Jupiter is sitting relative to us.

To tackle this problem, astronomers measure distances compared to the Earth's distance from the Sun. They call this unit an *astronomical unit*, or au, and by its definition, the Earth is on average 1au away from the Sun. Jupiter's distance can then be written as 5.2au, telling you that it is just over five times as far from the Sun as the Earth's position.

These distances are important because the distance from the Sun controls the type of dust that will be used to build the planets. Heated by its young star, the protoplanetary disc is much hotter near its centre than far from the Sun's rays. This temperature gradient determines which elements are able to condense into solids. In the same way that water becomes ice at 0°C (32°F), other molecules change from gas to solid dust particles at lower or higher temperatures. Closer to the Sun than even Mercury's orbit, the temperature exceeds 2,000°C (3,600°F) and evaporates all solids to form a region free of dust. As we step outwards, the temperature drops to 1,500°C (2,700°F), and the first dust particles form from metals such as iron, nickel and aluminium. At the Earth's orbit of 1au, silicates join in the mix, and when the temperature drops below freezing, ices appear. The first ice to solidify is pure water, made from hydrogen and oxygen. As the temperature cools still further, other hydrogen–based ices form, including solid methane and ammonia. These ices comprise much more common elements than the metals of the inner disc, leading to a burst of extra material where they can solidify. The point where ices appear is commonly referred to as the *ice line*, *frost line* or *snow line*, and it separates the terrestrial planets such as Earth and Mars from the gas giants like Jupiter. What is more, it helps explain their key differences.

Assembling from dust grains in the disc, each planet will consist of the solids that surrounded it as it formed. In the case of Mercury, this leads to a body predominantly made from iron.[*] After adjusting for its small size, which makes gravity squeeze it less than the Earth, Mercury's heavy material gives it a density that is the highest in the Solar System. As more molecules join in the mix of available dust particles, the density of the planets further from the Sun drops slightly but they remain similarly rocky. However, as we hit the ice line, the disc is swamped by low–density ices. With this extra boon of material, bigger objects can form that will one day become the cores of giant planets.

---

[*] Mercury turns out to be even more dominated by iron than expected from its hot location. The planet was probably involved in a collision that stripped away part of its non-iron crust, to leave it even more iron dominated than before. However, even this does not entirely explain its composition, which remains an open question.

However, while this fits into the picture of a planet assembled from its local dust grains, it doesn't explain how the connection process actually works.

The glue stick

Suspended in the gas, dust grains are more easily led astray than a child in a sweet shop. This is actually a good thing for planet formation, since if the dust remained on militaristic circular orbits then collisions would happen rarely and large objects would never form. It is fortunate for us that the dust has a wild side that bumps grains off their circular orbit and into the path of other grains.

This type of deviant motion was first observed in 1827 by a botanist named Robert Brown, who was studying pollen grains suspended in water. Brown noticed that the grains appeared to be moving randomly, but he could not decipher what was causing this motion. It took until the turn of the following century for the problem to be untangled by Albert Einstein, who recognised that the water molecules were bumping about the pollen. Einstein might have won a Nobel Prize for this discovery since it confirmed the existence of atoms and molecules, but he had already scooped one five years earlier for an entirely different project. The award instead went to the French physicist Jean Baptiste Perrin, in 1926, who experimentally confirmed Einstein's explanation. While Robert Brown's observations were insufficient for him to get an award, the effect was named after him and became known as *Brownian motion*.

In the protoplanetary disc, the gas plays the part of the water molecules that buffet about the small dust grains. In addition to the Brownian motion, dust grains are moved into collision courses by the gas's own slightly non-circular motion, which is encouraged by the magnetic field threading through the disc. Small pockets of slightly higher-density gas can also give a weak gravitational tug on the easily influenced tiny grains.

At the absolute beginning of the planet-assembly process, the sticking force between two colliding grains is less of a mystery. The dust grains that have condensed in the proto-planetary disc are less than a tenth of the size of sand grains, at only micrometres in size. Moving at speeds below 1m/s, these grains can be loosely held together by the electric charge of their atoms.

A dust grain consists of molecules such as ice or silicate, which are neutral with no overall positive or negative electric charge. Each of these molecules is made up of two or more atoms, which contain a central positively charged nucleus surrounded by equally negatively charged electrons. However, the electrons are not stationary. Instead, they scoot around the molecule and cause whichever side they briefly cluster at to gain a slight negative charge, while on the opposite side the molecule becomes positive. The charge from the negative end of the molecule can attract the positive end of a neigh-bouring molecule, holding the two together. This force from the slight asymmetry in electric charge is know as the *van der Waals force*, named after the Dutch scientist Johannes Diderik van der Waals. The force itself is actually fairly weak and can only work while the collision between dust grains is very gentle. When that is not the case, we begin (metaphorically and literally) to hit problems.

At micrometre sizes, the initial dust grains' random motion is slow enough that the van der Waals forces are sufficient to stick colliding dust together. The problem is that as the dust particles grow, so does their collision speed. Once the micrometre-sized dust grains reach the princely size of a millimetre, the van der Waals forces fail to provide a sufficient amount of stick. Instead, grains hitting one another bounce.

When two dust grains bounce, neither increases in size. The grains therefore increase from micrometre to millimetre sizes and then get stuck, forming a sea of millimetre particles.

This is a disappointing dead end to the planet-formation process unless by some chance a few dust grains make it to the centimetre scale. Experiments conducted in the laboratory have shown that when the two particles colliding have sufficiently different sizes, the smaller grain will rebound but leave up to half its mass behind. This is akin to throwing a jelly at your brother. A large amount of the jelly may fall to the floor, but there will still be a satisfying amount sticking to his face. Centimetre-sized dust grains could therefore collide with the pool of millimetre dust, gaining mass with each interaction.

While this is promising, it does leave unresolved the question of how to first get centimetre-sized dust grains. In fact, there are two methods that can be used to bypass the bouncing barrier. The first is dumb luck. While the average collision speed between the dust grains increases with their size, there is still a range of values. This makes it possible for a handful of collisions to have velocities sufficiently low that the van der Waals forces can build a centimetre-dust grain. The second option is that the bouncing barrier may be much less of a problem if you are fluffy.

Imagine throwing a rubber ball at a wall. If your aim is any good, it will bounce straight off the wall and smack you on the nose. Now imagine that the wall is replaced by a giant ball of dust and fluff, of the sort that lurks under sofas. The ball you throw will penetrate such a dust ball rather than rebounding. If the dust ball is big enough, your ball will remain stuck inside its fluffy interior and become part of its structure.

Protoplanetary dust grains may not consist of dust mixed with cat hair and fluff, but in the low gravity of space they can be fluffy. This is especially true for those made of lighter elements such as ice. Collisions between such fluffy particles are difficult to perform in the laboratory, since Earth's gravity will compress the grains. To get around this, the collisions can be performed virtually using computer simulations. This modelled reality showed that collisions between micrometre icy grains could stick, rather than bounce, for speeds of up to

60km/s (38mi/s). If the grains were still fluffy but made from silicates (as is more likely around the point where Earth forms), then sticking would still be effective up to 6km/s (3.8mi/s).

This seems like the solution to all our planet-formation problems. The micrometre slow-moving dust grains stick together via the weak van der Waals electric forces to form millimetre grains. The fluffiest of these aggregates can then stick together to form centimetre sizes, whereupon both fluffy and compact grains can gain mass during collisions with smaller grains. If this goes on for a few million years, Itokawa-sized objects can be formed that hold themselves together by gravity.

This would be the perfect solution if it were not for the gas disc.

🪐

Travelling in their orbit around the young Sun, the gas and solid particles feel different forces. For the smallest dust grains below a centimetre in size, this difference does not matter. Tiny grains are suspended in the gas, which carries them along like a baby in a sling, synchronising their velocities. As the dust grains grow in size to larger solids, they become more like hand-held toddlers. They are still orbiting the star, but their motion is no longer tightly linked to the surrounding gas. This is a problem because grains are solid while the gas is a fluid, and a fluid feels pressure.

In the absence of the gas disc, solid objects feel the tug from the Sun's gravity and the reverse supporting force from their own rotation. Their resulting motion is said to be *Keplerian*, a term named after Johannes Kepler, who described this orbit in his laws of planetary motion. The gas, meanwhile, feels these two forces and an additional pressure force. The pressure originates because the accretion of material on to the Sun makes the disc denser near the centre. While the solids are not affected, this gradient produces an extra outward force on the gas that makes it flow about 0.5 per cent slower than the Keplerian speed. The result is that the solids feel a headwind in the same way a cyclist does, with the slower gas pushing against their motion. And like a cyclist in a strong wind, the solids start to lose speed.

As the solids' speed drops, their rotation can no longer support them against the gravitational pull of the Sun and they begin to spiral inwards. This happens fastest for dust grains that have grown near to a metre in size, and can cause these boulders to hit the star within a few hundred years when starting from the Earth's position. The only way to completely avoid this is to get bigger.

Anyone who has to clutch their stomach while taking a flight in a propeller plane knows that small aircraft are much more susceptible to air turbulence than a large commercial Boeing 747. This is because the drag from the surrounding air currents has a much bigger effect when your mass is small compared to the size of your surface. Likewise, once the dust agglomerates into kilometre-sized objects, it is no longer greatly bothered by the drag from the gas headwind. Unfortunately, the 100 years it would take for a metre-sized boulder to crash and burn into the Sun is a great deal shorter than the time needed for it to grow by collisions into a drag-oblivious, kilometre-sized rock. This problem is referred to as the *metre-size barrier*, and to stop it removing all our planet-building work, we need a way to pause this infall.

When competing in a bicycle race, a team of cyclists groups together to form a peloton, in order to mitigate against the exhausting force of the wind drag. An athlete cycling alone must battle against the wind while biking along the route, but when behind other cyclists, he is shielded and expends much less energy. Cyclists in a peloton take turns battling at the front of the formation, often allowing their champion cyclist to conserve strength for the final stretch near the rear of the formation.

A protoplanetary version of the cycling peloton is behind an idea called the *streaming instability*. The concept is that the solid boulders' doomed walk towards the Sun would be paused if the gas drag could be switched off. Somewhat like in the cycling peloton analogy, this can be achieved by having enough solids in one place.

The spiralling path of the boulders as they move inwards through the disc is inevitably not going to be a homogenous affair. Knocked about by the gas, points along the route end up with higher concentrations of rocks. These clusters act like a peloton, reducing the headwind from the gas within that region. As new boulders are dragged inwards from further out in the disc, they hit the peloton and also slow as the gas drag is reduced. This adds to the peloton members and further reduces the headwind. The result is a runaway process, as the larger peloton is able to more easily collect incoming boulders.

Computer models of the streaming instability suggest that this protoplanetary peloton could gather together solids amounting to a total size of a few tens to a few hundred kilometres, comparable to the size of the dwarf planet Ceres. At this point, complicated gluing finally becomes unnecessary. Gathered together in the protoplanetary peloton is enough material to enable gravity to act and pull the rocks together in kilometre-scale objects. These solids now have a sufficiently respectable size that we can call them *planetesimals*.

In constructing the record-breaking tower, school students in Delaware went from bricks around 1cm (⅖in) in length to a tower 1,000 times larger. It was undeniably impressive, but this change in scale was outdone by the Solar System. In building a planetesimal out of dust, the protoplanetary disc assembled creations 1,000,000,000 times larger than its starting pieces. What is more, it is not yet done. We have now let gravity loose.

Gravity: the power tool

Getting gravity involved in the planet-building process is the equivalent of switching a pot of weak, school-safe glue for power tools. Planetesimals are now jostled in their orbits by the gravitational tug of the neighbouring rocks, which causes their paths to cross and the planetesimals to collide. While smaller bodies may still shatter or rebound from these impacts, their speed is not great enough to escape the gravitational pull of the largest planetesimals and they are pulled back in. The largest objects begin to eat everything in their path.

How fast a planetesimal grows hinges on the number of rocks it hits and adds to its mass. In the same way that a snowplough is more effective with a large shovel than a small one, a planetesimal will gather more material if it is bigger. As smaller planetesimals merge into larger bodies and increase in size, this rocky sweep-up becomes increasingly efficient, until the density of the small objects begins to drop. It sounds delightfully successful, but as it stands, it is not fast enough.

At the Earth's position of 1au, it would take 20 million years for a planetesimal to snowplough into enough rocks to become our planet. If we take into account the sweep-up becoming less productive as the number of surrounding rocks thins, this time increases still further to 100 million years. As we move further from the Sun, the planetesimals become more spread out and their density drops. By Jupiter's distance, 100 million years becomes the minimum time in which the giant planet's solid central core can be formed. This is longer than the life of the gas disc, which must persist until the core has formed in order to source Jupiter's massive atmosphere. Once we reach Neptune, the planet core would need longer than the life of the Solar System to gather its mass. This means that we need to give this growth process a speed kick.

Fortunately, gravity's sticking power does not begin at an object's surface. While gravity's attractive force weakens away from the planetesimal, it can still drag nearby objects into a collision course. This results in an enhanced effective size for the planetesimal, equal to its geometric size plus a booster factor from gravity's reach. This boost is proportional to the planetesimal's mass, and therefore increases along with the geometric area as the planetesimal gets larger. The process becomes so efficient that the speed at which the planetesimal sweeps up new material gets faster with its size, leading to an ever-increasing growth rate. In this runaway phase, the biggest planetesimals rapidly accrete their surrounding neighbours. It is the planet-building version of 'the rich get richer'.

A planetesimal in runaway growth mode might continue to expand until it has eaten the entire disc, if it was not for the central star. A nearby small planetesimal passing close to a larger body will feel two forces: the gravitational pull of its neighbouring massive planetesimal and also that of the star that it is orbiting. The point where these two forces balance is known as the *Hill radius* of the massive planetesimal. Inside the sphere of this radius, the planetesimal's gravity is a stronger force than the pull from the star.

Since even a planetesimal in runaway growth mode will be vastly smaller than the star, the Hill radius is close to the planetesimal compared with the star's distance, although potentially many times the rock's size. Anything within the Hill radius will be pulled on to a collision course with the runaway planetesimal, but objects outside will still feel a tug. In fact, a planetesimal cannot keep to a safely stable orbit unless it is more than about three-and-a-half times the Hill's radius from its neighbour. Once jostled from its orbit, the planetesimal's path can cross the Hill radius and be devoured. A growing planetesimal can therefore feed from a path around 7 Hill radii in width as it orbits the star.

As the planetesimal grows, its Hill radius also expands to give a larger feeding zone, within which smaller planetesimals can be attracted. When the planetesimal and its Hill radius are small, the objects it accretes are those on close orbits. However, as the planetesimal goes through runaway growth, its increased Hill radius allows it to pull in bodies from a much wider area of the disc. These objects are initially moving at significantly different speeds from the main planetesimal, and are hauled off their orbits by its gravitational pull. Due to the strength of this force, these smaller planetesimals hurtle towards the central attractor at much higher speeds. This allows them to resist being channelled into a direct collision and instead they loop around the main planetesimal in chaotic orbits. This is much less efficient than everything neatly colliding. As a result, the runaway growth

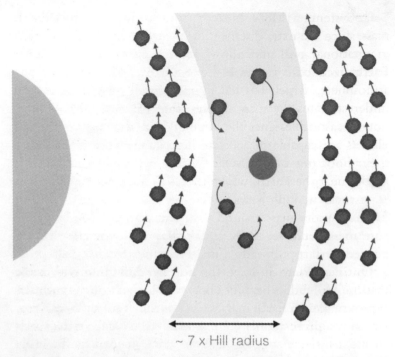

~ 7 x Hill radius

Figure 6 Feeding zone. A growing planetesimal can feed on smaller planetesimals within roughly 3.5 x Hill radius as it orbits. The Hill radius is the region where the planetesimal's gravity dominates over that of the star.

slows and a new phase of *oligarchic growth* (from the Greek, 'rule of the few') ensues.

Oligarchic growth still allows the largest planetesimals to grow, but slower than their slightly smaller neighbours in the runaway growth phase. This results in a catch-up process, where smaller objects grow faster than the most massive ones.

As the number of small objects decreases, the fresh food supply entering the planetesimal's expanding Hill radius dries up and the growth eventually stops. At this stage the planetesimal reaches a maximum mass known as the *isolation mass*, whereby it has eaten all the other objects in its orbital path. With a path width of around 7 Hill radii, the isolation mass is about 10 per cent of an Earth mass for an object at our 1au position, based on the estimate of the available mass from our

Solar System's MMSN. Near Jupiter, this increases to 1 Earth mass, since the extra distance from the Sun weakens our star's gravitational pull and allows for a larger Hill radius. An Earth-sized core is not big enough to reel in a large gas atmosphere, which has led to speculation that the MMSN underestimates the mass in our giant planet region. This is not unreasonable, since the huge gravitational pulls of the big planets are capable of accelerating planetesimals so fast that they shoot out of the Solar System entirely. This is much harder near the Earth, where the Sun's gravitational grip can keep rocks making a dramatic exit. If a young solar system did have more mass around the giant planets, then a typical core mass could reach around 10 Earth masses; the amount needed to start pulling in a massive atmosphere.

Out near Pluto at 40au, the pull from the Sun is so weak that the Hill radius becomes huge, giving an isolation mass of approximately 5 Earth masses. This is much larger than Pluto, which weighs in at only 0.2 per cent of the Earth's mass. Such an inconsistency suggests that the time needed for Pluto to clear its orbit is still longer than the age of the Solar System. While even today it is true that Pluto remains embedded in a sea of smaller objects (a fact that led to it being designated a *dwarf planet* in 2006), the comparison with its isolation mass is not entirely fair, since these distant objects probably did not form in their current positions.

Our planet-forming planetesimals are now dubbed *planetary embryos*, and somewhere between 30 and 50 of them would have sat between the orbits of Mercury and Mars. While initially forming on different orbits, the paths of the embryos do not remain separate. Collisions occur both with one another and with fresh planetesimals scattered in from the outer Solar System. In a violent, gladiator-style end game, the collection of embryos merges to leave just four terrestrial worlds.

To trigger such giant impacts, a massive gravitational bully is needed to scatter the path of embryos and planetesimals – and for that, the planetary embryos beyond the ice line need to become the gas giants.

CHAPTER THREE

The Problem with Gas

While a collection of large rocks travelling around a star may be a good basis for planet formation, this stony ensemble cannot yet be compared to planets. What is needed is for these planetary bases to acquire a topping of atmosphere.

At the beginning of the assembly process, our planets were dust-sized specks at the mercy of the gas disc. Carried by the gas motion, these solid grains had to avoid a catastrophic meeting with the Sun until they were able to grow big enough to resist the gas drag.

As they now approach the size of planetary embryos, these roles reverse and gas can become trapped by the embryo's gravitational pull. Once bound to the embryo's vicinity, the gas forms an envelope around the rocky core to create the planet's first primitive atmosphere.

Like the solid particles in the disc, the gas has a randomly directed motion in addition to its orbital path around the Sun. The speed of this movement in the gas is governed by the temperature. This is what makes a hot-air balloon inflate; warming the gas increases its speed and causes the molecules to bump harder against the fabric to expand the balloon. When the speed needed to escape the gravity of the planetary embryo exceeds the gas's random motion, the gas becomes trapped to form an atmosphere.

The acquisition of an atmosphere helps the growing planetary embryo sweep up smaller planetesimals. Such rocks entering the atmosphere feel a drag like the air resistance on a skydiver. As it loses speed, the planetesimal is more easily pulled towards the embryo's surface and collides to add to its core.

The braking of the incoming planetesimals also releases heat. In the same way that meteors become burning-hot shooting stars as they enter the Earth's atmosphere, descending planetesimals heat up as they are forced to slow. This energy warms the atmosphere and increases the random motions of the gas. While this boost in speed is not usually enough to allow the atmosphere to escape the planetary embryo, it does stabilise the gas against the embryo's gravity and stops it from being compressed. When gravity is balanced by the gas heat, the atmosphere neither expands nor contracts. This stable position is referred to as *hydrostatic equilibrium*.

As the planetary embryo grows through planetesimal accretion, its gravity increases. This temporarily breaks the hydrostatic equilibrium and pulls the atmosphere more tightly to the planet. As it is compressed, the gas heats up and balances the gravity again to make a new stable position. With the embryo's stronger gravity extending its reach and the gas compressed to make more space, new gas can pour in to deepen the atmosphere.

For a planetary embryo with about a tenth of the Earth's mass (as expected at the Earth's current position at this time), the captured atmosphere is always much smaller than the planet's solid mass. The gaseous topping therefore helps the embryo's growth, but does not drastically change the planet's evolution. However, as we step away from the Sun and cross the ice line, the story becomes very different.

The giants of gas

Further away from the Sun's gravitational pull, the larger cores around Jupiter's current location can trap much bigger atmospheres – so big, in fact, that the weight of the gas can never be supported by the heat from the incoming planetesimals.

At exactly what point this occurs is debated. A rough rule of thumb suggests everything can stay balanced until the atmosphere reaches the same mass as the embryo's solid core. However, less mass is needed if the incoming planetesimals partially evaporate in the atmosphere as they descend towards the embryo's surface, depositing their vaporised material into the gas. The heavier elements that make up the planetesimal ice and rock are coolants, quickly reducing the temperature of the gas, which slows its motion and weakens the support against gravity.

When this critical atmosphere size is reached, a balance point between the gas motion and gravity cannot be found. Instead, the combined mass of the embryo and atmosphere creates a gravitational force that overwhelms the gas motion. Hydrostatic equilibrium is shattered and the atmosphere is steadily compressed.

As the atmosphere packs down close to the planet, the embryo's gravity is able to pull in fresh gas from the disc. This joins the atmosphere and also begins to compress. The new gas adds to the embryo's combined mass, increasing its gravitational reach and allowing it to pull still more gas into the atmosphere. This produces another runaway process whereby the planetary embryo's atmosphere is accrued at a faster and faster rate. The result is a massive atmosphere of thousands of kilometres; a gas giant planet has been born.

There are two ways in which this massive atmosphere accretion can be stopped. In one way the atmosphere can continue to grow until the gas disc disappears. Once the star begins to evaporate the disc, the pool of gas surrounding the planet steadily reduces. Before 10 million years are up, the disc vanishes to leave the planets with whatever atmosphere they have managed to gather.

This method is certainly effective, since a planet cannot accumulate atmosphere if its gas reservoir has gone. It probably played the leading role for the outermost gas giants in our Solar System. Forming so far from the Sun, Uranus and Neptune will have had only a low density of rocks and gas to feed from, making the creation of their planetary embryos a slow process. It is therefore likely that they were still gathering their atmospheres when the Sun evaporated the remains of the gas.

As a small addendum, Uranus and Neptune are in fact *so* far out, that they probably did not form in exactly their present locations. Due to the length of time it would have taken to produce a planet of their size, the gas disc would have vanished before they could acquire a good atmosphere. It is more likely that they formed closer to Jupiter and Saturn and later moved outwards. Nevertheless, even at their estimated closer location, the evaporation of the gas disc probably terminated their atmosphere growth.

However, the above method is less probable for the inner two giants, Jupiter and Saturn. Our two most massive gas behemoths are thought to have a much larger ratio between their solid cores and huge atmospheres. It is therefore likely that they had plenty of time to acquire gas and that a different mechanism came into play to stop their feeding. This mechanism is suspected to be a gap that formed in the protoplanetary gas disc along each planet's orbit.

When in orbit, the distance away from the star determines the length of time it takes to complete one circuit. As happens on lanes on a circular running track, material closest to the star has the shortest distance to travel to return to its original position. Gas orbiting between the planet and the star therefore draws ahead of the planet, while gas further out lags behind.

The gravity of the planet pulls on the gas as it moves through the disc. For gas closer to the star that is trying to move ahead, this tug pulls it back and slows it down. Conversely, the outer gas feels a pull to speed it up.

As the gas speed changes, it must alter its orbit so that its circular speed can once again balance the star's gravity. Gas between the planet and star now has a lower circular speed and is forced to pull away from the planet and move closer towards the star. Meanwhile, the accelerated outer gas is able to move further away. This produces a gap around the planet where the gas density is much lower.

If the gravitational reach of the planet pierces the top and bottom of the protoplanetary disc, this gap can remain. The planet is so big that gas is unable to sneak into the hole without its speed changing and forcing it back out. The gap therefore persists and throttles the gas flow until the gas disc evaporates.

After its atmosphere stops growing, the planet will contract as its current atmosphere continues to cool and collapse. This causes the atmosphere to get denser, whereupon it gets harder to compress and begins to resist further shrinking. Deep inside a gas giant's atmosphere, the compressed gas reaches high enough pressures to transform its hydrogen gas into an exotic liquid metal. These incredible forces reduce the contraction of Jupiter and Saturn to an extremely slow rate, with current estimates for Jupiter hovering at less than a millimetre a year. This minute shrinkage is still enough to heat the planet, which radiates more energy than it receives from the Sun.

This mechanism for forming a gas giant planet is known as the *core accretion model*, a term inspired by the image of the runaway gas accreting on to the solid core. One of the most compelling features about it is that the method is very similar to terrestrial planet formation, with the only difference being a runaway atmosphere gathering. However, its least compelling feature is the time it takes.

Initial estimates for how long it would take Jupiter, Saturn, Uranus and Neptune to form in their current orbits are in excess of 10 million years; a problem since the gas disc will be gone within that time. At one point, this was thought to make core accretion impossible, but since then some tweaks to the system have been discovered that can shorten the time needed.

The first adjustment is simply to improve the accuracy of the original model calculations. How fast the gas cools is partially controlled by whether dust grains in the atmosphere clump and sink, or if they remain suspended in the gas. In the latter case, the fog of grains prevents heat from escaping (technically referred to as increasing the atmosphere's opacity), and slows cooling. Allowing the dust to fall towards the core gives cooling a boost and the atmosphere swiftly collapses into a runaway mode.

A more energetic solution is to move the planet. The planetary tow truck is the same mechanism that eventually causes the gap opening in the disc. While the planet pulls on the gas to attempt to open the gap, the gas in turn is tugging back on the planet. The inner gas tries to pull the planet forwards as the planet drags it back, while the outer gas drags on the planet as it is pulled forwards. If both the inner and outer gas pull equally, the planet will remain unaffected. However, the planet moves slightly faster than the local gas as it is not susceptible to pressure. This causes the dragging gas to be closer to the planet, dominating the inner accelerating forces. The planet therefore slows and gets an inward push.

As the planet moves through the disc, it enters a new population of planetesimals. This fresh supply of food allows its accretion rate to increase again, reducing the time needed to begin runaway collapse by an impressive factor of 10. In this scenario, a planet like Jupiter would begin to form further out at around 8au, then begin a food trawl towards its current position at 5au. Upon the discovery of exoplanets, this concept of planet migration became a key component in formation theories: as both an asset to the formation process and one of its biggest thorns.

More recently, another mechanism has been suggested to accelerate the weight gain of the gas giants. Rather than consuming large planetesimals, *pebble accretion theory* suggests eating smaller rocks may be more fattening.

Planetary embryo growth slows once the incoming planetesimals can move fast enough to escape the embryo's gravity. The difficulties begin at the oligarchic growth stage, and become worse for capturing larger planetesimals that are later scattered into the embryo's neighbourhood.

However, even after larger planetesimals have formed in the disc, there is still a large population of smaller rocks. At around 10cm (4in) across, pebbles are an excellent-sized snack because they are still affected by gas drag. The drag slows the pebbles as they swing by the embryo, making it far easier to divert them into a collision course. Embryos can therefore accrete this size of rock extremely efficiently, gaining mass a hundred times faster at Jupiter's current position.

In practice, all three of these mechanisms are likely to occur to shorten the time needed to pile on a huge atmosphere. This makes core accretion the preferred option for the formation of the majority of gas giant planets. Yet, some worlds would still challenge its limits.

Building distant planets

Despite core accretion being a serious player for giant planet formation, not everyone was happy. In particular, the further out in the disc you went, the harder it became to form a planet. For small rocky outer worlds such as Pluto, their far-flung positions can be blamed on interactions with the massive planets. As the gas giants balloon in size, their gravitational reach pulls first large planetesimals, then smaller rocky embryos towards their core. Too large to be affected by gas drag, the majority of these objects are moving too fast by the time they reach the gas giant to be captured. Instead, they accelerate past the planet to shoot all over the Solar System.

Pluto was shot outwards with a large collection of dwarf planets and planetesimals to sit beyond Neptune. Other planetesimals were scattered inwards or out of the Solar System entirely. So large was the gravitational tug from the mammoth Jupiter, that the embryos located in the inner Solar System were knocked about and collided to form the terrestrial planets.

This could just about explain our Solar System: planetary embryos formed via successful collisions of dust grains to create planetesimals. The gas giants acquired their huge atmospheres in a runaway process during core accretion, and their massive bulk triggered a game of gravitational ping-pong to complete the inner planets' growth and push out a ring of dwarf planets and rocks. Then, we discovered the exoplanets.

Fomalhaut b is thought to be a giant planet orbiting its star at a staggering distance of 119au. In comparison with our own Solar System, our outermost gas planet, Neptune, is as a measly 30au from the Sun. At hundreds of au, it is not possible to build up a large enough core to pull in a massive atmosphere, yet Fomalhaut b weighs in with an upper mass estimate three times that of Jupiter. This hit the core accretion model hard, both due to Fomalhaut b's location in the sparse outer disc, and by tripling the target mass that needed to be reached.

Fomalhaut b was not discovered by either the radial velocity or transit technique. Instead, it was the first exoplanet to be seen directly. Direct imaging of exoplanets is extremely difficult, since their dim radiation (from reflected starlight and heat) is usually overwhelmed by the star. The further away a planet orbits, the better the chance of spotting its faint signature.

The planetary nature of Fomalhaut b has been questioned, since it is surrounded by a large dust cloud. Is a planet buried in the fog, or is this the rubbly ruins from planet-making collisions? Either way, Fomalhaut b was not the last object to be found far out in the disc.

In 2009, the 8.2m (27ft) Japanese telescope *Subaru* began combing the skies for far-out planets. The survey was dubbed the 'Strategic Explorations of Exoplanets and Disks with Subaru', or *SEEDS* for short. The plan was to directly image discs around stars and any visible gas giant planets. By 2016, SEEDS had found four planets significantly larger than Jupiter and orbiting between 29au and 55au away from their star. These outer planets might not be numerous, but they were here to stay.

With the current model for core accretion pushed back beyond its limits, scientists hunted for an alternative gas giant factory. The proposal was to more closely mimic the way stars are formed.

Images of disc galaxies similar to our own Milky Way display a dazzling collection of spiral arms. The spiral is usually a compression wave, which is of the same type as sound. These spiral arms can appear when the gas's own gravity is strong enough to break apart the smooth structure of the disc.

Such an effect is known as a *disc instability*: a rather fancy term for saying that gravity breaks the disc apart. In the resultant spiral arm, natal gas clouds are gathered together to create more of the dense pockets where stars are born.

The idea behind the second form of gas giant formation is that the protoplanetary disc could behave in the same way. Spiral arms would develop in the gas disc encircling the star, and these would compress the gas, which would then collapse directly into a giant planet. Unlike the much lower densities commonly found in the natal star-forming cloud, densities in the protoplanetary disc could potentially rise high enough to form a small, planet-sized object.

It is hard not to be enticed by such a prospect, since it neatly avoids every other problem we have been trying to deal with. With no need to first build up a solid core, sticking mechanisms and the gas drag on planetesimals can be negated. The formation time for the gas giant can now be as little as a thousand years, a fraction of that needed for the core accretion model and well within the lifetime of the gaseous protoplanetary disc. Moreover, it should be quite possible to create planets of 1–10 Jupiter masses, encompassing Fomalhaut b and other exoplanet mega-worlds.

The problem (and there is always a problem) is that it is questionable whether a protoplanetary disc can form instabilities in the same way as a Galaxy. There are two main factors that decide if a disc can become unstable: mass and temperature. If the disc is too light, its gravity is not sufficient to upset the smooth distribution and form a spiral. Conversely, if the temperature is too hot, the random motion of the gas can swiftly smooth out the compression wave before it can form a planet. It is also not certain if planets formed in this manner can survive. Closely forming planets may merge into bigger objects or they may shred each other apart.

Models of a protoplanetary disc encircling a star like our own Sun suggest that anything inside 40au is very unlikely to become unstable. However, in the disc's younger and heavier days, it could have fragmented beyond 100au: a figure that fits well with Fomalhaut b's location. For the planets found by SEEDS, their location puts them on the cusp between where core accretion and disc instability seem to be possible. Their formation mechanism is therefore perhaps understood, but the jury is out as to which method we should apply.

The gaseous planet formed via a disc instability initially has no solid core. It can acquire one by capturing planetesimals that slowly sink to the planet's centre. While our own gas giants are too close to the Sun to have formed via disc instabilities, the core for a Jupiter-sized planet that was formed via this mechanism is around 6 Earth masses; within the guess range for our own Jupiter's solid centre.

So between the two theories of core accretion and disc instability, which is correct or could both be at work? The only real reason to reject the idea that both mechanisms occur is aesthetics: it is just ugly to have two different gas giant formation methods. Yet, neither model alone can explain the gas giants in our own Solar System and match those around other stars. A compromise might be that the two methods complement and help one another: disc instabilities compress gas in spiral waves, which in the right circumstances can collapse into a planet. Where collapse does not happen, any disc instabilities may still assist core accretion, as the gathered gas boosts the rate at which planets can gather atmosphere around a solid core.

At this point, our planets are becoming recognisable as the worlds of our Solar System. The four outer worlds formed the fastest, gathering in planetesimals and smaller embryos from an ever-increasing gravitational reach. With this extra bulk, huge atmospheres poured down to drench the rock and ice cores in gas. In the inner Solar System, the stronger pull from the Sun restricted the gravitational influence of the planets, causing formation to proceed more sedately. The gas giants' gravity then ruffled the orbits of the inner planetary embryos to begin a final set of collisions. Out of these impacts, four terrestrial planets would emerge, coated with thin atmospheres. However, not a single one of these planets is yet capable of supporting life.

CHAPTER FOUR

Air and Sea

'Oh, my God! Look at that picture over there! There's the Earth coming up. Wow, isn't that pretty!'

The comment was made by William 'Bill' Anders aboard NASA's *Apollo 8* spacecraft, while undertaking the first manned mission to orbit the Moon. As Anders would later tell space historian Andrew Chaikin, 'We came all this way to discover the Moon. And what we really did discover is Earth.' His photograph of *Earthrise* was an unplanned part of the voyage's itinerary, but it became one of the most iconic images in history.

At the end of 1968, the US Central Intelligence Agency (CIA) had gleaned evidence that the USSR was on the brink of sending two of its cosmonauts on a mission to loop around the Moon. If successful, this would be the first journey to take humans beyond Earth's orbit, and a significant step in the space race to put a person on a different world.

NASA had launched its first crewed mission of the Apollo programme earlier that year. *Apollo 7* had carried astronauts Walter Schirra, Donn Eisele and Walter Cunningham around the Earth 163 times over 11 days. Despite the crew coming down with colds, resulting in congestion that was difficult to ease in the low-gravity environment, all operations went smoothly. Plans for *Apollo 8* were rapidly assembled, but America was not yet ready for the historic Moon landing that would ultimately see astronauts Neil Armstrong and Buzz Aldrin walk on the Moon's surface in 1969.

NASA had good reason to be worried that it would be beaten to the punch. While the exact state of the Soviet space programme was unknown, a couple of Russian tortoises had already made the Moon loop, returning to Earth in September that year in the first mission to circle the Moon and land back on Earth. The risk that the USSR might beat the US to its goal was very real.

NASA therefore put forward a daring proposition: to send *Apollo 8* on a round trip that would orbit the Moon and then return to Earth. This had risks since *Apollo 8* would be launched on the *Saturn V* rocket, whose previous unmanned test flights had suffered from serious problems with vibration. Yet, those were thought to be fixed and this was not the moment for doubt. Before his assassination, President John F. Kennedy had committed America to besting the USSR in the race to the Moon. Second place would be last.

On Christmas Eve that year, the three astronauts Frank Borman, James Lovell and Bill Anders became the first humans to see the dark side of the Moon with their own eyes. Yet, it was not the pockmarked landscape of our natural satellite that captivated their attention.

As *Apollo 8* rounded the Moon, a blue–white Christmas ornament rose on the desolate horizon. It was our own planet, rising over the lunar surface. Anders's *Earthrise* snapshot became – as the late nature photographer Galen Rowell declared – 'The most influential environment photograph ever taken'. It was a pale blue dot and it was ours.

A second atmosphere

Had Anders been able to photograph the Earth some 4 billion years earlier, he would have captured a vision of hell. Riddled with heat from its violent formation and still pounded by planetesimal collisions, our planet was molten rock with volcanoes spewing oceans of magma. Appropriately, this first epoch in the Earth's history is known as the *Hadean* period, named after Hades, the Greek mythological god of the Underworld.

The initial cloak of gases the Earth had gathered from the protoplanetary disc was not to last. Consisting mainly of light elements such as hydrogen and helium, our planet's gravity was too weak to hold securely on to this first atmosphere. As the Sun went through its violent T-Tauri adolescence, winds and radiation poured from the young star and ripped away the gases from the inner terrestrial worlds.

Stripped of its primitive atmosphere, the molten Earth began to spew a second set of gases. The mobility of the melted rock allowed denser metals to sink towards the core, leaving lighter silicates to form the mantle. Gases previously absorbed by the rock were released as it melted, escaping in volcanic plumes to form the Earth's second atmosphere. The air was now a mix of water vapour, carbon monoxide and carbon dioxide, and nitrogen. Free oxygen was still absent and would not appear until the beginnings of photosynthesising life. The molecules in the new air were heavy enough to be grasped by the Earth's gravity and avoid being stripped. It might not yet be the atmosphere we breathe today, but it was the prequel to life.

Yet, in this atmosphere lay a mystery. Where did the water vapour belched by the Hadean Earth come from?

The mystery of water

The *Earthrise* photograph shows a delicate blue planet with 71 per cent of its surface covered by water. By mass, this fraction is far smaller, with the surface water and the estimated quantity tucked into the mantle amounting to less than 0.1 per cent of the entire Earth. But wherever water is found on Earth, so is life. From our point of view, this means that it's a key ingredient.

The conundrum is that because of the Earth's proximity to the Sun it was born out of dust too warm to contain ice. Instead, the solid particles that collided to build the Earth consisted mainly of dry silicates. Ice could not solidify into grains to join the planet–formation process until we stepped out past the ice line and into the chilly neighbourhood of the gas giants.

The effect of this can still be seen in rocky members of the asteroid belt: a band of leftover planetesimals that sits between Mars and Jupiter, right around the ice line's location. Since collisions and interactions between the asteroids can send members skittering towards the Earth to fall as meteorites, we have a reasonable amount of information about their properties. What has been found is that asteroids that sit in the outer part of the belt at 2.4–4au contain a sizeable amount of ice, enough to comprise to 10 per cent of their mass. Inwards through the belt towards Mars and the Earth, the asteroids dry out, with only about 0.05–0.1 per cent of their mass in ice. While the passing of billions of years has altered the ice fraction, this trend suggests that our planet was formed out of dry material. But if that is the case, where did our oceans come from?

Wet Earth

Option one is known as the *wet Earth* scenario. In this picture, the forming Earth is able to gather water as a gas.

While no solid ice could form in the inner Solar System, the protoplanetary disc was abundant in water vapour. This bathed the planetesimals as they collided to build the Earth. If enough vapour could cling to the rocks, water could have been incorporated into the planet itself. As the Hadean Earth's molten interior shifted, the water vapour could escape through the volcanoes and eventually cool to form the oceans.

A similar alternative is that the Earth was able to hold onto water vapour gathered in its primitive atmosphere. While the light hydrogen and helium were stripped, the heavier water molecules were retained to join the later secondary gases.

A wet Earth formation is a definite possibility, but it does have a few unresolved problems. The issue with retaining gas from the protoplanetary disc is the abundance of noble gases; elements such as helium, neon and argon that are particularly unreactive. The lack of chemical action makes it hard to change the amount of these gases over time. If part of our atmosphere came from the disc, then the fraction of noble gases in the Earth's atmosphere should be similar to that in the Sun. In fact, they are much less abundant in our air, suggesting that our atmosphere was outgassed and not captured. Moreover, to capture enough water in the atmosphere, the

Earth's primitive atmosphere must have been very large. However, the Earth's formation was reasonably slow and took longer than the lifetime of the protoplanetary disc. There might not have been enough time to reel in sufficient gas. This would not prevent water from being incorporated within the planetesimals, but the issues are enough to suggest water might have been delivered after the Earth formed.

Dry Earth

Water-delivery services come in the form of ice-laden meteorites. Originating in the outer regions of our Solar System, these rocky packages could have formed in ice before travelling towards the terrestrial planet neighbourhood. The Moon's cratered surface is evidence of the heavy bombardment of rocks that must have been raining down on the terrestrial planets during their formation. While the Earth's atmosphere vaporised many incoming meteorites and its surface was renewed and smoothed by volcanic activity, the Moon retains the pockmarked visage from its pummelled teenage years. If our planet had formed dry, an incessant pummelling of icy material could have given us our oceans.

The influx of these rocks can largely be blamed on the gas giant planets. Their huge mass created a gravitational pull that sent any local leftover planetesimals catapulting around the Solar System in a game of gravitational pinball. Since these rocky pinballs were formed in the gas giant neighbourhood beyond the ice line, they were packed with ice when they came shooting into the inner Solar System and on to the terrestrial planet worlds.

This theory for water delivery still leaves us with unknowns. In particular, there are multiple pockets in the Solar System that contain the scattered remains of the gas giant's pinball game, each with a different history. Finding the one populated with rocks akin to those that delivered our oceans would reveal a lot about how the Earth became habitable, and in turn provide the prospect of finding a second planet that could support life.

Just past the edge of our planets lies one such band of these discarded pinballs commonly known as the *Kuiper belt*. Situated at 30–50au, this collection of rocky bodies circles the Sun just beyond Neptune. The most famous member of the Kuiper belt is the dwarf planet Pluto, but it is thought to be accompanied by around 100,000 other sizeable objects that each have a bulk exceeding 100km (62mi).

The Kuiper belt's name gives credit to the Dutch-born American astronomer Gerard Kuiper, who proposed that these objects could have formed in the early Solar System. However, this naming is somewhat controversial, since Kuiper's paper in 1951 is pre-dated by eight years with a similar prediction from Irish astronomer Kenneth Edgeworth. Kuiper actually thought that this band of objects would no longer exist, believing that a massive Pluto would have scattered away all the other members. He therefore un-predicted the group that bears his name. In fact, Pluto's mass is much smaller than Kuiper believed, and has very little influence over its Kuiper belt companions. For these reasons, the band of objects is sometimes referred to as the *Edgeworth-Kuiper belt* to acknowledge both independent predictions, or *trans-Neptunian objects* by people preferring irrefutable accuracy over questionable credentials.

The formation mechanism of the Kuiper belt does remain uncertain. It is not impossible that this collection of objects could have formed at its current location, but its large distance from the Sun presents challenges. So far out, the dust grains in the protoplanetary disc would be spread out along their wide orbit. This lowers the chance of the collisional encounters needed to build the 100km- (62mi) and 1,000km- (621mi) sized bodies. The problem is exacerbated by Neptune, whose gravitational pull ruffles the inner Kuiper belt region, increasing the speed of its material. The faster-moving grains and planetesimals then struggle to collide slowly enough to stick, reducing their rate of growth still further. This interference can be avoided if the inner Kuiper belt objects could form before Neptune, but this puts even greater pressure on how fast they must gather material.

It is therefore more likely that these rocky bodies were scattered outwards by the gravity of the outermost gas giants, Uranus and Neptune, allowing the planetesimals and dwarf planets to form in the more collision-rich environment closer to the Sun, before being booted into the outskirts of the planetary system. Neptune certainly has a strong connection with the evolution of the Kuiper belt. Its largest moon, Triton, is an ex-Kuiper belt object that Neptune stole for one its own satellites. Unlike most other moons in our Solar System, Triton orbits Neptune in the opposite direction to the planet's rotation and has a similar composition to Pluto; strong indications that it did not form alongside Neptune but joined its orbit later.

Neptune's interactions with the Kuiper belt are not limited only to their formation. If the rocky members come too close to Neptune's bulk, they can be accelerated once again to shoot towards the inner Solar System. As the Kuiper belt object approaches the Sun, its icy body begins to vaporise and it grows a tail of water vapour. It has become a comet.

Named after the Greek word for *long hair*, a comet appears in the sky as a fuzzy globe of light at the head of an extended tail. Some comets are on long orbits around the Sun, reappearing in the sky every few decades or centuries. Others pass by our planet only once, leaving the Solar System permanently after a single dance past the Sun.

The sudden appearance of comets amid the regular constellations has meant they have been associated with auspicious or foreboding events throughout history. Halley's Comet is a particularly famous spectacle. On an orbit taking 75–76 years, it has been immortalised in the 70m- (230ft) long Bayeux tapestry, created in the 1070s to depict the Norman conquest of England. The same comet's appearance in 1301 inspired the Florentine painter Giotto di Bondone to depict a comet as the star that led the wise men to the birth-place of

Jesus in his religious fresco, the *Adoration of the Magi.*[*] The comet is named after the British astronomer Edmond Halley, who was the first to connect sightings of a comet in 1456, 1531, 1607 and 1682 with the same object on its periodic orbit. Halley predicted that the comet would reappear in 1758. He did not live to see this occur, but the arrival of the comet on schedule led to it bearing his name. Halley's Comet last appeared in our skies in 1986 and will make its next visit in the middle of 2061.

The association of Halley's Comet with a mysterious cosmic portent is perhaps appropriate, since the origins of this comet are similarly disguised. With an orbit taking less than two centuries, Halley's Comet is considered (rather ironically) a *short-period* comet. This class of comet usually comes from the Kuiper belt, kicked out during a close encounter with Neptune to shoot towards the inner Solar System. The gravitational kicking results in the orbits of these comets being highly elliptical, unlike the near-circular orbits of the planets. However, since both Neptune and the Kuiper belt sit roughly in the same disc-shaped piece of space as the original protoplanetary disc, the comet their interaction creates also moves around the Sun in that plane.

But this is not true for Halley's Comet. The orbit of Halley's Comet is inclined at such a large angle compared to the planets, that it actually flips the comet around so that it orbits the Sun in the opposite direction. This means that while most short-period comets only modestly rise out of the disc's plane by less than 10 degrees, Halley's Comet is on an orbit inclined at an angle of 162 degrees. The strange motion is an indication that the comet began its journey in another location. That location is the *Oort cloud* at the knife edge of the Solar System.

<hr>

[*] Halley's Comet would have been visible around 11BC, but the timing is not quite right for this to have truly been the biblical signpost to Bethlehem.

While the Kuiper belt objects may have been tapped outwards by the planets, the Oort cloud is filled with rocks that received a far more substantial boot. Approaching close enough to the gas giants to be strongly accelerated by their bulk, these rocks were slingshotted to the very edge of our Solar System. At this critical point, the Sun's inward gravitational pull is balanced by the outward tug from the gravity of the rest of the Galaxy. This creates a precariously stable region where these rocks do not feel a pull in any direction and so could come to rest; a knife edge at the brink of our solar system.

This tenuous spot for rocky collectables is so far out that the Oort cloud has never been directly observed. Its distance from the Sun is estimated at between 22,000 and 100,000au (over 1 light year) with a population of rocks that is speculated to extend into the trillions.

Balanced as they are between the Sun and the Galaxy, a gentle prod from the gravity of a passing star can topple rocks in the Oort cloud like bottles perched on a wall. A rock will then shoot towards the inner Solar System to be caught in the Sun's gravity as a long-period comet.

A more recent visitor than Halley's Comet to our neck of the solar neighbourhood has been Comet Lovejoy. Discovered by Australian amateur astronomer Terry Lovejoy in August 2014, the comet became visible to the naked eye (no telescope required) at the start of 2015 and made its closest approach to the Sun in late January that year. Unlike Halley's Comet, Comet Lovejoy is on an incredibly long loop around the Solar System, with a period that was initially set to be 11,000 years. After passing through the planet-populated part of the Solar System, the gravitational tugs from the planets changed its orbital path to be a shortened (yet still staggeringly long) 8,000 years. By contrast, Pluto's orbit is a mere 248 years in length.

By virtue of the fact that it is a loop, an object on an orbit must return to its original starting location. This means that comets on orbits longer than 200 years must have begun further away than the Kuiper belt; the bent elliptical shape of their path taking them far past Pluto. This fact, coupled with the varied inclination of these long-period comets, was what caused Jan Hendrik Oort to propose the existence of a shell of distant objects surrounding the Solar System.

Somewhat ironically, not only were Kuiper and Oort both Dutch astronomers, but their postulations about the Kuiper belt and Oort cloud were both pre-dated by someone else's work. Oort's proposal for the origin of the long-period comets was made in 1950, but was previously suggested in 1932 by the Estonian astronomer Ernst Öpik. Öpik's suggestion that the long-period comets originated in a cloud far beyond Pluto was published in the *Proceedings of the American Academy of Arts and Sciences* – a location that fellow astronomer Fred Whipple noted in an article about Öpik's work was 'a journal not much searched by astronomers for astronomical contributions'. Öpik's paper was also not helped by its nondescript title, 'Note on stellar perturbations of nearby parabolic orbits'; a substantially less eye-catching start than that of Oort's 1950 publication, 'The structure of the cloud of comets surrounding the Solar System'. Oort himself apparently overlooked this earlier work, since he credits Whipple in his paper for drawing his attention to Öpik's contribution, saying:

> I am indebted to Dr Whipple for drawing my attention to the interesting article by Öpik in which also the action of stars on a cloud of meteors or comets is discussed. This article, which I have only been able to read after the first three sections of the present paper had been written, deals with the influence of passing stars on greatly elongated orbits.

This casual style is a contrast to today's scientific journals, where a detailed comparison with Öpik's model would doubtless have been required, irrespective of how many sections of the paper had already been written!

While Öpik remains less associated with the idea than Oort, the Oort cloud is sometimes referred to as the *Öpik-Oort cloud* to acknowledge both the ideas.

For both reservoirs of comets, Halley's Comet remains an anomaly. It has a period too short to reach the Oort cloud, but an inclination too high to come from the Kuiper belt. In fact, while the period of Halley's Comet has been approximately the same since at least 260BC, it was probably once much longer and shortened by interactions with the planets. This situation where a comet from the Oort cloud has a period comparable with the Kuiper belt comets is unusual but not unique. To date, just under 100 Halley-type comets have been discovered, compared with the total comet count of more than 5,000.

The comet reservoirs form the first delivery-service option for putting water on Earth. Forming in the gas giant neighbourhood, comets are packed with ice. During the same scattering processes that landed them in their current homes, many similar rocks would have flown inwards through the Solar System, able to crash down on Earth's dry surface. Yet, was this incoming delivery the one that gave us our oceans?

The answer lies in the comets themselves. If they provided our water billions of years ago, then the water that has solidified into their icy body should resemble the water found on Earth.

Surprisingly, not all water is identical. One of the most important variations is the ratio of the amount of hydrogen to its heavier sibling, deuterium. Both hydrogen and deuterium are simple atoms with a single electron. They differ in their central nuclei, with the hydrogen atom containing a single proton while deuterium contains both a proton and a neutron. A water molecule can consist of an oxygen atom bonded to two hydrogen atoms, or it can have one or both of its hydrogen atoms replaced with a deuterium atom. In the latter case, it becomes known as 'heavy water' (or 'semi-heavy water' if the molecule has one hydrogen and one deuterium atom) to reflect the weight of the extra neutron. Heavy water occurs naturally on Earth, but only in small quantities. Roughly one in 6,700 atoms of hydrogen on Earth is deuterium. If this ratio matches that found in comets, then they might well be the source of our seas.

The best way to find out the water content of the comets is to catch one. This was achieved by one of the most ambitious space missions of the decade: Rosetta.

The International Rosetta Mission was launched by the European Space Agency (ESA) in March 2004. The aim was to chase down Comet 67P/Churyumov-Gerasimenko, follow this icy rock as it passed the Sun and land a probe on its surface. Previous space missions had flown close to comets before, but none had orbited a comet nucleus or been brash enough to attempt to park directly on the comet itself.

In fact, Comet 67P/Churyumov-Gerasimenko was not the first choice for Rosetta, which was originally set to head to Comet 46P/Wirtanen. However, a failure during a launch of two communication satellites at the end of 2002 caused the mission to be delayed by a year to investigate the issues with the *Ariane 5* rocket. This pause in the proccedings caused the comet to move beyond reach and in May 2003, Comet 67P/Churyumov-Gerasimenko was selected as the new target.

Comet 67P/Churyumov-Gerasimenko takes the second part of its name from its discoverers, the Ukrainian astronomers Klim Ivanovych Churyumov and Svetlana Ivanovna Gerasimenko. The discovery happened by serendipity, when Churyumov was examining a photograph taken by Gerasimenko of another comet, 32P/Comas Sola, on 20 September 1969. On examining the image closely, Churyumov realised that there was a second object in the frame, which turned out to be a previously undiscovered comet.

Since the double-barrelled name of these two astronomers is a mouthful to pronounce, the comet is often referred to by the first part of its moniker, '67P'. This cipher tells you that the comet was the 67th periodic comet to be discovered.

Unsurprisingly, given its extensive history, the first periodic comet on that list is Halley's Comet, which holds the full designation 1P/Halley.

With an orbital path that takes about six-and-a-half years, Comet 67P is a short-period comet that is thought to have once been part of the Kuiper belt. Up until a few centuries ago, the comet only came within 4au of the Sun; a distance just inside the orbit of Jupiter that was too chilly to vaporise the icy body into a tail. This meant that the dormant rock was unobservable from the Earth. Then, in 1840, its path changed. As Jupiter and the comet both orbited the Sun, a chance close encounter caused the gas giant's gravitational pull to drag on the comet and alter its orbit. This happened again in 1959, reducing the comet's closest approach to one slightly larger than the Earth–Sun distance at 1.29au, resulting in its detection 10 years later. Due to Jupiter's influence on Comet 67P it also gained the classification of a *Jupiter Family Comet*; a class of objects whose orbits are controlled by the giant planet.

The Rosetta mission was named after the famed Ancient Egyptian *Rosetta Stone*. Now in the British Museum in London, the stone is inscribed with a decree issued on behalf of 13-year-old Ptolemy V in 196BC. It was written after a revolt in the city of Lycopolis, during which the temple priests refused to pay taxes to the pharaoh, was successfully subdued. The decree affirms the royal status of the young ruler and goes as far as to declare him a god to be worshipped in all temples across Egypt.

The relevance of the Rosetta Stone is not in its proclamation of heavenly attributes, but in the fact that it is repeated in three different languages. At the top of the stone, the decree is written in Egyptian hieroglyphics, a script suitable for such important pronouncements. However, underneath this, the decree is transcribed again in the Egyptian demotic script for everyday use, and again in ancient Greek, which was widely used for administration. The appearance of the same message in three writing systems was a key factor in deciphering the hieroglyphs; a devilishly difficult task due to the mix of phonetic and pictorial characters. It was this that inspired the spacecraft's name; while the Rosetta Stone provided a window into the secrets of the hieroglyphs, the *Rosetta* spacecraft was aiming to uncover the secrets of the comets.

Rosetta caught up with the comet while it was between Mars and Jupiter, at roughly 3au from the Sun. It had been a circuitous trip, requiring three loops around the Earth and one around Mars to receive kicks from the planets' gravity to help the spacecraft reach such a distant location. In total, *Rosetta* travelled 6.4 billion kilometres (4 billion miles) on a journey that took more than a decade to complete. As the distance from the Sun increased and solar energy became in short supply, *Rosetta* went into hibernation. It awoke the January before its autumn meeting with the comet in 2014 with a cheerful 'Hello, World!' announcement on its Twitter feed, although amusingly, the social media platform had not existed when *Rosetta* left the Earth.

While the majority of the science programme would be completed by *Rosetta*, it was the descent of the lander to the comet's surface that captured the world's imagination. The size of a refrigerator, *Rosetta's* robotic lander was named *Philae* after the Egyptian island on which a discovered obelisk provided additional clues for decoding the hieroglyphs. Its descent to the comet's surface on 12 November 2014 took seven tense, nail-biting hours. It was tension shared worldwide. Despite the risks of a public failure, the ESA broadcast the event live over the Web to 10 million viewers. At 16.02 GMT, *Philae* was confirmed to be on the comet's surface and the world celebrated. Unfortunately, the lander was not to stay there.

Due to a double failure with both *Philae's* thrusters and harpoons, the small probe was unable to anchor itself to the comet's surface. Instead, it bounced; a dangerous event that could have potentially thrown the lander free of the comet's low gravity. Miraculously, *Philae* did return to the surface, but came to rest on unstable terrain in a region overcast with shadow. The shade prevented *Philae* from charging its solar-powered secondary batteries. As a result, it went into hibernation after draining its primary battery in two-and-a-half days.

There was initial hope that the lander might reawaken as the comet approached the Sun. However, apart from a short burst of intermittent communications in June 2015, the lander remained silent. In February 2016, the ESA declared that it had given up hope of hearing from *Philae* again. Despite its short duration of activity, however, *Philae* managed to complete 80 per cent of the initially planned science programme. Meanwhile above the comet, *Rosetta* continued to take data.

With a gravitational field several hundred thousand times weaker than the Earth's, *Rosetta* could not go into a true free-fall orbit around the comet. Instead, it used thrusters to make a triangular motion about the nucleus that steadily brought it inwards to a minimum of 10km (6.2mi) from the comet's surface.

As *Rosetta* dipped within 100km (62mi) of the rocky terrain, it plunged through the nebulous coma of gas that engulfed the comet's nucleus. This was the first taste of the comet's water. It was not like Earth's. Comet 67P had water that was more than three times more deuterium-rich than Earth's oceans.

While *Rosetta* had been the first spacecraft to travel with a comet on its trip around the Sun, a much earlier ESA mission had flown past Halley's Comet in March 1986. The *Giotto* probe had sampled the comet's coma and its measurement was joined by 10 other ground-based observations for different comets flying in from the Kuiper belt and Oort Cloud. Only one of these had a heavy water fraction similar to that on Earth.

The data from *Rosetta* would form hundreds of journal publications, but the result of the comet's water analysis was one of the earliest. Published in the journal *Science*, in December 2014, the conclusion was that water gets heavier as you travel further away from the Sun. This pointed to an origin closer to home for Earth's seas.

Situated between Mars and Jupiter, the asteroid belt is our closest collection of rocky leftovers. Somewhere between one million and two million objects larger than 1km (0.6mi) in size orbit the Sun in a band that is a bit more than 1au in width. When NASA's *Pioneer 10* probe made the first journey through the asteroid belt in July 1972, there were concerns that a chance collision with one of the many rocks could destroy the spacecraft. In fact, the vast region of the belt ensures that the asteroids are widely spaced, with average separations of a few million kilometres.

Like the Kuiper belt, the asteroid belt has its own dwarf planet, Ceres. It also has a number of notable rocks, including the asteroids Vesta, Pallas and Hygeia, all larger than 400km (250mi) across.

The asteroid belt might have been the location of another planet, but Jupiter's looming gravitational influence increased the collision speeds of the rocks to make it difficult for a new world to be formed. Like Neptune on the far side of the gas giant neighbourhood, Jupiter also scattered rocks in and out of the region. The asteroids on the Jupiter side of the belt are therefore more water-rich, forming just past the ice line and supplemented by rocks bumped inwards by Jupiter.

An unfortunately timed tug from Jupiter's gravity or a collision between belt members can occasionally knock asteroids from their orbits to send them skittering towards the Sun. Unlike comets, asteroids do not contain enough ice to grow tails. Instead, their new location in the Earth's neighbourhood gains them the designation *Near Earth Object*, or NEO.

Although the paths taken by a few NEOs can make them a threat to Earth, their closer passage does have advantages. In particular, that they are much easier to visit. This was the plan for two spacecraft launched in 2014 and 2016: the Japanese *Hayabusa2* and the American *OSIRIS-REx* missions.

Hayabusa2 is the successor to the mission that visited the asteroid Itokawa in Chapter 1. Itokawa is an *S-type*, or *stony*, asteroid; a class of space rock that typically originates in the inner part of the asteroid belt. This location makes S-type asteroids dry and victims of space weathering, where their surfaces are battered and changed due to the barrage of solar wind particles and radiation. Such evolution means that while Itokawa revealed much about the lives of asteroids, it could not uncover the secrets of the rocks that struck the early Earth.

For this reason, *Hayabusa2* is chasing down a different type of asteroid. Its target is a rock known as *Ryugu*; a *C-type*, or *carbonaceous*, asteroid, which is thought to have changed relatively little since the formation of the Solar System 4.56 billion years ago. While Ryugu now orbits the Sun between the Earth and Mars, it probably began life with the main collection of C-type asteroids at the far icy side of the asteroid belt.

Hayabusa2 was launched at the start of December in 2014 and is due to reach Ryugu in 2018. It is hotly pursued by the *Origins, Spectral Interpretation, Resource Identification, Security, Regolith Explorer (OSIRIS-REx)*, launched by NASA in the autumn of 2016 and headed for a second C-type asteroid, *Bennu*. Both spacecraft are on sample-retrieval missions. This means that they will not only transmit information about the asteroids back to Earth, but will actually return home with grains gathered from the asteroids themselves. As *Philae* has already proved, the landing required to gather the samples is an immensely difficult task, but the potential gains are high.

Meteorite finds on Earth that show evidence of once containing water are also packed with organic molecules. These point to the tantalising possibility that the crash landing of rocks on the early Earth did not just bring water, but the starting kit for life itself. Samples from Ryugu and Bennu will therefore not test just the hypothesis for the origin of the Earth's oceans, but the very beginnings of Earth's life.

Asteroid Ryugu's name comes from a Japanese folk tale about a fisherman named Urashima Taro. In the tale, Urashima rescues a sea turtle being tortured by children. Somewhat fortuitously, this particular turtle turns out to be the daughter of the Emperor of the Sea. As a reward for his kindness, Urashima visits the underwater palace of Ryugu-jo and spends three days with the princess in human form. However, on his return home, Urashima finds that 300 years have elapsed. In confusion, he opens a box given to him by the princess. This releases a cloud of smoke that clears to leave Urashima an ancient man. The box had contained his old age.

Hayabusa2 and *OSIRIS-REx* will return to Earth respectively in 2020 and 2023. It is hoped that they will be carrying boxes that contains the secrets to the Earth's habitability, in the same way that Urashima's box from Ryugu contained his life.

PART TWO
DANGEROUS PLANETS

The Impossible Planet

CHAPTER FIVE

The Impossible Planet

For planet-formation theories, the discovery of 51 Pegasi b was rather unfortunate.

Its announcement in 1995 as the first planet found orbiting a star like our own Sun threw open the door to a new era of planet discoveries. It also blasted a hole in the theory of planet formation.

In truth, 51 Pegasi b was not the first exoplanet to be discovered. Around five years earlier, a planet had been found orbiting the remnant of a dead star known as a pulsar. However, since a pulsar is very unlike our Sun, there were any number of excuses for why such a planetary system should be different from our own. For 51 Pegasi b, the excuses were harder to find. This was a planet orbiting a Sun-like star but in completely the wrong place.

51 Pegasi b is a gas giant planet. It has a minimum mass approximately half that of Jupiter, making it 150 times more massive than the Earth. However, it sits so close to its star that one year on 51 Pegasi b is over in a blisteringly short 4.2 days. Even our Sun's closest planet, Mercury, takes 88 days to complete an orbit, while our Jupiter takes a full 12 Earth years to loop the Sun.

Such differences from our own Solar System were intriguing, but they resulted in a conundrum: both the major theories for gas giant planet formation required the planet to form far away from the star.

To gather enough mass to capture the colossal atmosphere of a gas giant, the planet must form beyond the ice line so that it can bulk up on frozen ices. It also needs to be far away from the star so that its gravity can dominate over a wide region and lure in a large reservoir of planetesimals (in the language of Chapter 2, its Hill radius needs to be big). For formation via disc instability, the planet needs to be even further out than the ice line, 51 Pegasi b was so close to its sun, that it should not have been able to become a gas giant. In fact, the intense heat at such a location should have prevented much solid mass from forming there at all.

Just to add insult to injury, 51 Pegasi b did not prove to be an isolated anomaly. As detections of exoplanets began to mount up, so did the number of gas giants that snuggled up to their stars.

It was true that current observational methods favoured finding such *hot Jupiters* over planets in systems that might resemble our own. Very massive and close to their star, these huge worlds caused the maximum wobble in the star's radial velocity. Their swift orbits also repeated the signal of their presence every few days. In short, hot Jupiters were the bawling babies of the exoplanet field. Nevertheless, there was no denying their presence. Later statistics would estimate that around 1 per cent of stars hosted a hot Jupiter. You could not have a planet formation theory without them.

There was only one logical explanation for their existence; if such a planet could not be formed where it sat, then it must have been born further away and moved there.

The concept of a planet changing its orbit was actually not a new one. The idea had been postulated as early as the 1980s, but had originally been dismissed. The problem had not been how to start a planet moving, but how to stop it.

Planetary migration occurs when the gravity of the growing planet begins to pull strongly on the surrounding gas in the protoplanetary disc. The gas resists, with the swifter gas closer to the star trying to pull the planet forwards, while the slower gas further out in the disc drags back on it. Since the planet does not feel the gas pressure, the directly surrounding gas is also part of the slower moving component. This makes the backward drag the stronger force, and the planet loses speed and moves inwards towards the star.

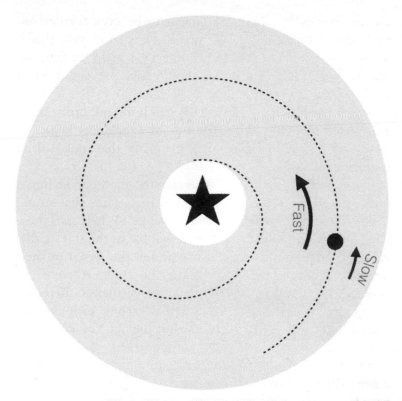

Figure 7 A planet migrating through the protoplanetary disc. Gas closer to the star moves faster than the planet and tries to pull it forwards. Gas further out is moving slower and drags it back. Typically, the drag wins and the planet slows and moves towards the star.

We touched on this motion when exploring gas giant formation in Chapter 3. Trawling through the protoplanetary disc was an effective way to grow by gathering planetesimals. Unfortunately, it can end very badly for the planet.

Predictions for the planet's inward migration suggest a merciless rate. Within 100,000 years, a gas giant embryo at Jupiter's current position could meet a fiery death as it crashed into the star. This is far shorter than the time taken for the disc to evaporate and release the planet from the drag of the gas. Since a planet's gravity can be sufficient to start migration once it reaches the size of Mars, it is unclear how it is possible for planets to form at all.

This is the second time on the planet factory production line that gas drag has nearly cast would-be new worlds into the star. The first occasion was the drag on small planetesimals, which feel a headwind in the slower-moving gas. As the planetesimals grow into embryos, this drag can no longer affect their more massive body. But as the mass increases further to form small planets, the planet's gravity pulls on the gas to create dangerously strong drag forces once again.

The obvious existence of our Solar System was the reason why the idea of migration was initially dismissed. Yet, the discovery of hot Jupiters gave reason to pause. Was it possible for a planet to migrate, but stop before it hit the star?

While the gas drag forces the planet to change orbit, the planet's pull does the same to the gas. The slower-moving gas is accelerated and forced to move outwards, while the faster-moving gas is slowed and moves inwards. This pushes the gas away from the planet. When the planet is small, fresh gas is able to pour in to replace the displaced material. However, when the planet's gravity becomes strong enough, it pushes all gas away to open a gap in the disc.

It is this process that completed gas giant formation described in Chapter 3; the young gas giant grows rapidly as it migrates into fresh planetesimal reservoirs. As its mass becomes great enough for its gravity to open a gap, the planet is left in a low-density hole and atmosphere growth finally stalls.

With the gas pushed away from the planet, its countering drag might entirely disappear. However, the disc gas is also drifting inwards as it is accreted by the star. This causes the gap to try and refill from the outside edge, providing a renewed supply of gas to drag back on the planet before it is pushed away. The result is still an inward push, but substantially weaker than before. If the planet can move slowly enough, the gas disc will evaporate and leave the planet free of its force.

Planetary movement before the gap opens is known as *Type I migration*, and progresses to *Type II* once the planet carves a hole. However, due to the speed of Type I migration, planets run the risk of never making it to the sedate Type II mode.

Exactly how Type I migration is stopped remains an open question. One possibility is that the inward-moving planet acts as a snow shovel, piling up gas just inside its orbit. This increases the fast inner gas that wants to pull the planet forwards, allowing it to counter the slower gas's drag. Sudden bumps and changes in the gas, such as those at the ice line, may also flip the strength of the inward and outward gas forces to act as *planet traps* and halt this mode of migration. In short, anything that can affect the flow of the gas in a region of the disc may also change the rate of Type I migration.

If hot Jupiters reached their current location via migration, then migration potentially plays a tight game with planet formation. Assuming the hot Jupiters we observe stopped their inward march when the gas disc evaporated, their proximity to the star suggests that they got very lucky. Alternatively, the planets may stop at the disc's inner edge, beyond which the star has evaporated or accreted all the material.

However, while migration offers a solution to the hot Jupiters, it messes up the Solar System.

The problem with Mars

If migration does indeed explain the hot Jupiters, we are left with an obvious question: how did the planets in our Solar System avoid the same fate?

Whether the terrestrial planets were hauled about by migration is a debated topic. Forming more slowly, the Earth and its neighbours may have stayed below the mass to drive fast migration until the gas has evaporated away. Alternatively, our rocky worlds may have been held in position by one of the planet traps mentioned earlier.

The fate of the gas giants is less easily waved away. The fast formation needed to ensure their huge atmospheres forces the planets to be susceptible to both Type I and Type II migration. Even if Type I migration could be slowed or halted, their huge mass would have opened the gas gap to begin Type II migration and sent the planet onwards towards the Sun.

We have also seen hints that at least a small amount of orbital movement did shape our planets. Gas giants can more easily gain mass if they migrate through the disc. The current positions of Uranus, Neptune and the Kuiper belt may also not have contained enough material for these objects to form, suggesting that they may have shuffled over from a denser region. But if migration did occur, what stopped Jupiter from crashing through the inner Solar System and destroying the Earth?

In fact, Jupiter may have tried to do exactly that. The clue to this dangerous past lies with Mars. Despite being named after the Roman mythological god of war, Mars is small and weedy. It is so diminutive that its size has presented a problem for planet-formation theories.

As we step outwards through the inner Solar System to Mercury, Venus, Earth and Mars, the pull from the Sun's gravity weakens. This allows the gravitational influence of the planet (its Hill radius) to extend and attract rocks from a wider area as it forms. The result of this progressively larger feeding zone ought to be steadily bigger planets. We should therefore find that planet masses increase until we approach Jupiter, whose huge gravity starts to mess with the formation process to create the asteroid belt.

This logic holds well until we pass the Earth, but rather than Mars being a super-sized version of our own planet, it has only a tenth of our mass. Even allowing for the proto-planetary disc to gradually drop in density between the Sun and the ice line, we would expect a Mars of between 0.5 and 1 Earth mass. Moreover, the asteroid belt should also be chunkier and populated with a collection of Mars-sized embryos. Instead, the largest object in the asteroid belt is Ceres, which is around a hundred times smaller than Mars.

We can fix this conundrum if the density of planetesimals was abruptly depleted at around the Earth's current position.

Deprived of planet-building blocks, both Mars and the asteroid belt are then forced to remain small. But what could have happened to drain Mars's neighbourhood of rocky supplies?

When searching for missing mass, an obvious suspect is the most mammoth planet in the Solar System: Jupiter. Could Jupiter have had a wild history that entailed the planet travelling into the inner Solar System, gathering or scattering planetesimals, then migrating backwards to its current location?

This idea became known as the *Grand Tack* model, named after the sailing manoeuvre to reverse the direction of a boat. The scenario begins when Jupiter is forming in the protoplanetary disc. As the planet's gravity starts to pull more strongly on the surrounding gas, the young Jupiter begins to migrate towards the Sun. The change in orbit allows the growing planet to quickly gather more planetesimals, eventually gaining enough mass to open a gap in the gas. This slows Jupiter's motion to Type II migration, but the planet keeps shrinking its orbit. Jupiter's march also shepherded planetesimals in its path inwards, while scattering others outwards. The planet might have ended up as a hot Jupiter, were it not for the appearance of Saturn.

Further out in the protoplanetary disc, Saturn formed more slowly than its big sibling. While Saturn is the second-largest planet in our Solar System, it has less than a third of the mass of Jupiter. Due to its lighter weight, Saturn opened only a partial gap in the gas, allowing the planet to migrate swiftly and gain on Jupiter's inward march.

As the planets approach, their years become closer in length. Eventually, Saturn orbits the Sun exactly twice in the time it takes Jupiter to orbit three times. This ratio is known as a 2:3 resonance, and it is very difficult to break.

Exactly why resonances are stable can be seen by picturing a 1:2 resonance between the two planets. Saturn would then orbit the Sun twice in the time it takes Jupiter to orbit once. For the first half of its orbit, Saturn is behind Jupiter and the larger planet's gravity pulls Saturn forwards. For the second

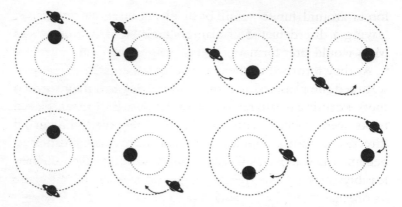

Figure 8 A 2:1 orbital resonance. The inner planet pulls the outer planet forwards for the first half of its orbit and backwards for the second half. The net force on the planets due to each other is zero, making a stable set-up that is difficult to change.

half of its orbit, Saturn is in front of Jupiter and being pulled back. Overall, these forces cancel out each other, and neither planet feels an extra tug during its circles about the Sun. However, if the two planets move slightly closer, the balance breaks. There is now an overall pull on Saturn that encourages the planet to gain speed and move outwards to the resonance position. This same balance applies to planets in different orbital resonance, such as 2:3 or 1:4.

Due to this stability, resonant orbits are common in planetary systems. Neptune orbits the Sun three times for every one orbit of Pluto, while Jupiter's moons Ganymede, Europa and Io are all in resonant orbits around the giant planet in a 1:2:4 ratio.

When Jupiter and Saturn reach their 2:3 resonance, they are close enough that the gap around Jupiter and partial gap around Saturn overlap. Instead of being tugged by gas on both sides, Jupiter now only feels the pull of the faster inward gas, while Saturn feels the drag of the outer gas. This causes Saturn to want to migrate inwards, while Jupiter now wants to migrate outwards. Since the resonant orbits push the planets apart if they try to move past one another, it comes down to a battle of strength. With the stronger gravity, the forces around Jupiter win. Both planets migrate outwards, leaving a disc depleted of planetesimals right around where Mars would later form.

As they returned to the outer Solar System, the planet pair scattered the planetesimals that had moved into their vacated spots. Forming in this region beyond the ice line, the displaced rocks were packed with ice. They were sent flying in all directions, with a population ending in the asteroid belt to become the water-rich C-type asteroids. Others flew further, feasibly striking the newly formed Earth and delivering its oceans.

The tack of Jupiter and Saturn also prevented Uranus and Neptune from drifting towards the Sun. As the smaller gas giants form and begin to migrate, they also risk becoming locked in resonant orbits. This makes it very difficult for the planets to jump past their big siblings and continue towards the Sun.

The Grand Tack model was proposed in 2011 in the journal *Nature*, led by astronomers Kevin Walsh and Alessandro Morbidelli. The day after seeing the success of the Grand Tack model in replicating both Mars's small size and the asteroids, Morbidelli strode into Walsh's office and declared that he had shook his finger at Jupiter the night before and told the giant planet, 'Jupiter! I know what you did!'

When Jupiter and Saturn near their present positions, the protoplanetary gas disc evaporates. The planets are finally free of the gas drag for good. However, to finish our Solar System's tale, we need one last reshuffle.

Exactly how this mayhem goes down is still debated. There are two main hypotheses: the *Nice Model* and (with a naming choice reminiscent of Hollywood blockbusters) the *Nice Model II*. In both scenarios, chaos erupts due to the planet factory leftovers.

Just beyond the giant gas planets lay a sea of remaining planetesimals. These rocky remains had skirted the edge of the planet orbits and avoided being absorbed or booted out of the Solar System. In the *Nice Model*, the lifting of the gas drag allowed the coercing pull of the giant planets' gravities to be a dominant force that causes these rocks to dribble inwards.

The strong gravity of the gas giants could rapidly accelerate nearby planetesimals. Moving too fast to be captured by the planet, these scattered out of the neighbourhood. The result of ejecting a planetesimal was a reverse kick on the planet like the recoil from the firing of a gun. Being very massive, this backlash did not strongly affect the gas giant. However, the impact could build up through multiple scatterings and eventually make the planet change its orbit. This new motion is known as *planetesimal-driven migration*.

Due to the difficulties with forming Uranus and Neptune in their current positions, it is suspected that the planets were much closer together when the gas disc dispersed. Jupiter's position could remain around 5au, but Neptune would be around 15au, rather than 30au, with Saturn and Uranus in between. The scattering of the planetesimals separated this close configuration, causing the planets to move apart.

As their orbits diverged, Jupiter and Saturn reached a second orbital resonance. However, this time the resonance did not lock the planets together. While approaching planets can force one another to maintain the resonant orbits, planets that are moving apart cannot stop their motion. Instead, passing the resonance produced a gravitational jolt that bent the orbits of Jupiter and Saturn into a more elliptical path.

These bent orbits pushed the two largest planet towards Uranus and Neptune. The two smaller giants were scattered outwards, barging into the outer reservoir of planetesimals. This resulted in a truly massive scattering, with the smaller rocks shooting all over the Solar System. Some of these rocks were pushed outwards to form the Kuiper belt, others bombarded the inner planets, and a population left the planet-populated region altogether to find a home in the Oort cloud.

The Nice Model was named after the French town where the idea was formed. The Nice Model II kept the name and proposed a similar scenario between the outer planetesimal rubble and the gas giants. In this version, the planetesimals did not have to dribble inwards and be scattered. Instead, the gravitational pull of the whole rocky debris field was enough to break the resonances between the giant planets and start the resulting chaos.

Dramatic as these models sound, there is evidence for such a giant scattering event occurring on the surface of our Moon. Examination of the lunar craters shows a spike in activity around 700 million years ago.

With the planetesimals scattered away, the orbits of the giant planets finally settle. Uranus and Neptune now sit at their current more distant location with the remains of the planetesimal sea pushed out by Neptune's motion to form the Kuiper belt.

The planet with the density of polystyrene

Migration provided a fast and efficient train line to slide Jupiter-sized planets close to their star. With the Solar System's formation now consistent with migration, this seemed the solution to this strange class of planet. Yet, as more planets were discovered, a population of hot Jupiters appeared that did not quite fit the gas migration picture.

At first glance, WASP-17b appeared to be a regular hot Jupiter. It was named for being the 17th discovery in a ground-based survey to hunt for transiting planets;[*] a project known as the *Wide Angle Search for Planets*, to give the insect-inspired acronym 'WASP'.

The planet was found circling a star in the constellation of Scorpius around 1,300 light years from Earth. With one orbit taking just 3.7 days and a radius of 1.5–2 times that of Jupiter, this was clearly another hot gaseous world.

However, closer inspection of WASP-17b revealed two surprises: first, the planet was immensely bloated. Despite being a super-Jupiter in size, radial velocity measurements

---

[*] Using telescopes on Earth, rather than in space.

revealed a mass equivalent to only 1.6 Saturns. This small mass but extreme size gave the planet an average density of 6–14 per cent of Jupiter and a few per cent that of Earth. It was a value so low that UK astrophysicist Coel Hellier remarked that the planet was only 'as dense as expanded polystyrene'.

The second rather surprising discovery was that it was orbiting backwards.

In our Solar System, the planets circle the Sun in the same direction as the Sun's own spin. We call such orbits *prograde* and they are the expected situation since the Sun, proto-planetary disc and planets are formed from the same core of rotating gas.

With everything turning the same way, migration should not cause the planet's orbit to reverse. Like being pulled into the centre of a whirlpool, a hot Jupiter that has been dragged inwards should still be orbiting in the same direction, albeit much closer to the star. Finding WASP-17b orbiting in the opposite direction to its star's spin therefore did not fit with the idea of gas-driven migration.

WASP-17b's opposing orbit is known as *retrograde*. While the Solar System's planets do not display such contrariness, the same it not true of the comets. Halley's Comet orbits the Sun in the reverse direction due to being given a hard kick by a planet that flipped over its orbit. Such a kick is a possible explanation for WASP-17b. While no other planets have been found orbiting its star, it is possible that it has a distant hidden sibling. The other possibility is that the sibling belongs to the star.

Roughly a third to half the stars in our Galaxy are binaries, with two (or sometimes more) stars orbiting one another. How strongly the gravitational pull from a stellar sibling can affect the forming planets depends heavily on its distance. WASP-17 does not have an obvious stellar companion, but it is possible that a nearby star was able to mess with its planet.

The method for stars interfering with each other's planetary children was described independently by Soviet scientist Michail Lidov in 1961, and Japanese astronomer Yoshihide Kozai in 1962. At the time, the strangeness of exoplanet orbits

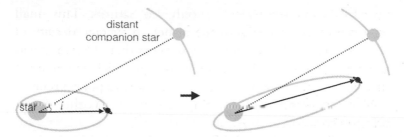

Figure 9 An edge-on image showing the Kozai-Lidov mechanism. A companion star (or more massive planet) can change the orbit of a planet by decreasing the angle (i) between the planet's orbit and companion's orbit and increasing the ellipticity.

was not known – Lidov was instead examining the orbits of moons and artificial satellites, while Kozai was looking at asteroids.

The pair of scientists studied systems where two large bodies are in orbit and one of these is also circled by a much smaller satellite. In Lidov's case, the two bodies were the Earth and Moon, and the small satellite was a space probe in the Earth's orbit. For Kozai's work, the two large bodies were Jupiter and the Sun, with the small body being an asteroid. They found that the second large body (the Moon or Jupiter) can perturb the orbit of the small satellite (space probe or asteroid). More specifically, the small satellite can lower its inclination (height) above the orbit of the two large bodies in exchange for increasing its orbit's ellipticity. This results in an alternating switch in the small satellite's orbit between height and ellipticity, where it will move from a highly inclined orbit to a highly elliptical one and back again. The mechanism became known as the *Kozai-Lidov mechanism*.

In the case of WASP-17b, the two large bodies of the Kozai-Lidov mechanism would be the planet's star and a second companion star, or even a more massive planet. Being the smallest object in the system, such a companion could start to change the orbit of WASP-17b.

This scenario allows WASP-17b to form in a tidy near-circular orbit beyond the ice line. As it begins to feel the tug from the second star's gravity, the height and ellipticity of its orbit begins to change. Eventually, the height can become so extreme that the planet flips over to follow a retrograde path.

As the orbit becomes more elliptical, its new bent path takes the planet closer to the star. This causes the gravitational pull from the star to increase during the nearby section of the planet's orbit, and decrease as the planet moves away. Varying the force from the star's gravity flexes the planet like a rubber ball. This produces heat, puffing up WASP-17b's atmosphere to exceed Jupiter proportions in an effect known as *tidal heating*. The energy to create the heat is removed from the planet's orbit, and the planet is forced to circle more closely to the star.* This fights the Kozai–Lidov mechanism, eventually forcing the planet into a close circular orbit. What is left is a hot Jupiter.

The Kozai–Lidov mechanism provides a second way of creating a hot Jupiter. But is a giant planet's migration more commonly due to gas drag, the pull of a distant star (or bigger planet) or the scattering from another planet?

In fact, all methods may be at work. For hot Jupiters in prograde orbits with no obvious star or massive planet companions, gas-driven migration probably dragged them inwards. Retrograde planets or those orbiting stars in binaries could have fallen victim to the Kozai–Lidov mechanism. The remaining hot Jupiters were then scattered inwards by another planet that either lurks further out, or was cast from the planetary system entirely when it kicked its neighbour towards the star.

This trio of options indicates that despite the surprise of hot Jupiters, it turns out there are many ways to create one.

As more planets were found around stars beyond our own Sun, other planets were found snuggled up to their stars. These new worlds were smaller than the hot Jupiters and unlike any planet we had yet seen.

---

* Like running around a valley, it takes less energy to run close to the bottom than to climb to the top and run around the rim. Likewise, widely spaced orbits have more energy than close ones.

CHAPTER SIX

We Are Not Normal

Twenty years after 51 Pegasi b shattered the theories of planet formation, astronomers came to an important conclusion: we are not normal.

Nearly 2,000 planets had now been discovered travelling around stars beyond our Sun. From these observations, it was estimated that 1 per cent of stars hosted a hot Jupiter, making these strangely located gas giants numerous, but still relatively rare. However, circling half of regular stars like our Sun was a type of planet unlike any seen in our own Solar System.

Labelled the *super Earths*, these planets were bigger than the Earth, but smaller than Neptune, with sizes stretching between 1.25 and 4 Earth radii. Most of the discoveries orbited their star in less than 100 days, with many circling even closer than the hot Jupiters. The most common orbit for the hot super Earths was at 0.05au; just 5 per cent of the distance from the Sun to the Earth, and 13 per cent of the distance to Mercury.

With a size between our largest rocky planet and our smallest gaseous one, what were these worlds? Had we discovered mega Earths with solid surfaces coated in thin atmospheres, or were these mini Neptunes, with small solid cores engulfed in gigantic gas envelopes? How did they end up so close to their star and why does our Solar System not have a planet that size? Could the answer be tied to how likely life is to arise in the Universe?

Without an analogue in our own Solar System, astronomers were left to untangle the origins of the most common type of planet in their data without having a familiar comparison.

In late 2011, NASA determined that after 34 years of travelling through space, its *Voyager 1* probe was on the brink of leaving the Solar System. The exact exit date was announced multiple times over the next few years to an increasingly amused public audience, spawning headlines that included 'Humanity leaves the Solar System – or maybe not' from *TIME* magazine, and 'Voyager has left the Solar System (this time for real!)' by the US NPR news. The problem was that defining the exact edge to the Solar System was a nearly impossible task, especially as we only had models to tell us what to expect to find there.

Despite these issues, the journey of *Voyager 1* made one fact absolutely clear: since it has taken from the probe's launch in 1977 until recent times to approach the boundary of our own planetary system, we were not going to visit an exoplanet any time soon.

Our nearest star (apart from the Sun) is Proxima Centauri; a dim star that sits 4.24 light years from Earth. The star is thought to host a planet with a minimum mass 30 per cent larger than Earth, making this our nearest possible exoplanet. Yet Proxima Centauri is still almost 2,000 times further than *Voyager 1* has currently travelled. At the space probe's current velocity of 60,000km/h (37,000mph), it would take more than 75,000 years to reach the nearest possible planetary system – so due to the vast distances involved, sending a probe to uncover the mysterious properties of the super Earths is a disappointingly unviable option. However, at least distinguishing between a rocky terrestrial world and a gaseous Neptune would be possible if we could measure the planet's density.

Born too close to their star's heat for ices to form, terrestrial planets like the Earth are built predominantly from silicates and irons. These heavy materials give these worlds high densities, with the values for Mercury, Venus, Earth and Mars ranging between 3.9 and $5.5g/cm^3$. For a similar composition, a higher mass planet will result in a higher density, as the stronger gravity acts to further squeeze the rock. Planet interior models reveal that a rocky super Earth with five times the mass of our planet would have a density of around $7.8g/cm^3$.

On the other hand, most of Neptune's bulk is in its huge atmosphere. This gas mainly consists of hydrogen and helium, the two lightest elements in the Universe. These dilute Neptune's density to be a low 1.6g/cm³. A gaseous version of the super Earth would be a mini Neptune, with a thick atmosphere engulfing a core of rock or ice. The density of a 5 Earth-mass gas planet might be around 3–4g/cm³; higher than Neptune due to its smaller mass attracting less light gas, but far below that of a rocky super Earth.

The average density of a planet is simply the planet's mass divided by the volume of space it fills. Since planets are approximately spherical, this comes down to two values: the planet mass and the planet radius. Unfortunately, not only is acquiring both measurements challenging, but the uncertainty in the recorded value can be large. For a planet finely balanced between a massive terrestrial world and a small gas giant, this uncertainty can leave us as much in the dark about the planet's type as would the sex of a foetus with its legs crossed in the womb.

One frustrating example of this was the super Earth orbiting the star Kepler-93. As its name suggests, Kepler-93 was observed by the Kepler Space Telescope in the search for transiting planets. The star was one of the brightest the telescope examined, allowing both a swift announcement of a closely orbiting planet in 2011, and an impressively precise measurement of the planet's size. Kepler-93b took only 4.7 days to circle its star and had a radius of 1.478 Earth radii, with an uncertainty of just 0.019 Earth radii, or 119km (74mi). This meant that the true value of Kepler-93b's radius lay in a narrow range between 1.459 and 1.497 Earth radii, designating it a clear super Earth.

This accurate radius measurement was followed up with an attempt to determine the planet's mass. Using the Keck telescopes on the summit of the dormant Hawaiian volcano of Mauna Kea, Kepler-93 was scrutinised for a telltale radial velocity wobble. The stellar jiggle was seen, but the motion was difficult to pin down. Initial estimates found that Kepler-93b weighed in at 2.6 Earth masses, but with a massive uncertainty that suggested the planet could be as large as 4.6 Earth masses. Further measurements helped to constrain the pattern of the star's motion, giving Kepler-93b a mass of 3.8 Earth masses, with an uncertainty of 1.5 Earth masses either up or down. This was better, but a mass range of 2.3–5.3 Earth masses inside a planet radius of 1.478 Earth radii left a large number of options for the planet type. In fact, the resulting average density range of 4–9g/cm^3 indicated that Kepler-93b could have been anything from a gaseous world through to a rocky terrestrial planet. So despite an extensive set of observations, the nature of Kepler-93b remained a mystery.

It was not until four years after the announcement of the radius measurement for Kepler-93b that the planet's nature was revealed. An additional set of radial velocity observations, using the European Telescopio Nazionale Galileo in the Canary Islands, nailed the planet mass down to 4.02 Earth masses, with an uncertainty of just 0.68 Earth masses. This yielded an average density estimate of around 6.88g/cm^3, pointing to Kepler-93b being a giant rocky planet. So did this mean that all super Earths were in fact larger versions of our own home planet?

In the beginning of 2014, astrophysicist David Kipping was searching for *exomoons*; moons orbiting extrasolar planets. It was an ambitious objective. The largest moon in our Solar System is Ganymede, a satellite of Jupiter with a mass double that of our own Moon and 2.5 per cent that of the Earth. While exoplanets might host larger moons, these planetary satellites would have only a tiny effect on the star.

A potential solution to this was not to search for the moon's effect on the star, but for its effect on a planet. Like the star and planet, a planet and its moon orbit a common centre of mass. This causes the planet to wobble during its orbit around the star. If the planet transits, the wobble will slightly alter the times of concurrent transits. The situation is similar to running circuits on an athletics track holding on to a small child. When the child is pulling you forwards you move slightly faster, and you slow when the child drags you back, causing a variation in your lap time. Observing a change in the time taken for a planet to appear back in front of its star between orbits might therefore be a smoking gun for the presence of a hidden moon.

This technique is known as *transit timing variations*, or TTV. In their 2014 journal paper, Kipping's team presented data from eight transiting planets, hunting for slight changes in their periodic appearances. To their excitement, they found a variation in one of the transit times. But this was not due to a moon.

The transiting planet was orbiting a cool star known as Kepler-138. Three planets had previously been identified by the Kepler Space Telescope, all with very small radii of between 0.4 and 1.6 of the radius of the Earth, and situated close to the star with orbits lasting less than one month. The wobbling planet was the outermost in the system, Kepler-138d. However, its wobble was not caused by unseen moon, but from the drag and pull of the middle neighbouring planet, Kepler-138c.

While it was initially disappointing that the first exomoon had not been discovered, the result still hit the record books. Like the wobble in the star, the variations in the transit time could act as a set of weighing scales for the planet's mass. Kepler-138d turned out to be the lightest world to have both its size and mass measured.* The previous record holder had been a rocky planet, Kepler-78b, which was 70 per cent heavier than the Earth. Kepler-138d clocked in at just 1 Earth mass.

Such a near match in mass to our home world ought to have made the nature of Kepler-138d obvious. This should be a rocky terrestrial world, too hot to host liquid water but with a solid surface and thin atmosphere. However, the radius of Kepler-138d was almost 60 per cent larger than the Earth's, making the density four times lower than that of our planet

---

* This was trumped by its own planetary sibling, Kepler-138b, whose 0.07 Earth masses was measured in 2015.

and only 30 per cent higher than water. This was no rocky world, but a very small Neptune.

Further observations in 2015 adjusted the mass of Kepler-138d downwards to 0.64 Earth masses, and the radius to 20 per cent larger than the Earth's. This still left the planet with an extremely low density of 2.1g/cm^3, and a thick atmosphere.

Speaking for the media, Kipping commented, 'This planet might have the same mass as Earth, but it is certainly not Earth-like. It proves that there is no clear dividing line between rocky worlds like Earth and fluffier planets like water worlds or gas giants.'

Our most common planet type would appear to be like a bag of mixed marbles: similar in size, but wildly different in design.

The varied nature of the super Earths had astronomers hooked. While Kepler-138d and Kepler-93b proved that there was no sharp divide between the massive terrestrial and small gaseous worlds, was there an approximate splitting point between the two planet types?

In 2014, roughly 70 super Earth planets had both mass and radius measurements. The average density of these planets suggested a rough rule of thumb that a planet with a radius of more than 1.5 Earths would have the thick atmosphere of a mini Neptune.

There were many exceptions to this rule in both directions. The size of Kepler-138d should have made it rocky, but it was gaseous. Meanwhile, a planet denoted BD+20594b was found to have a radius of 2.2 Earth radii, but with a density high enough to make it predominantly rock. Nevertheless, for cases where only the size of the planet was known, the 1.5 Earth-radius rule provided a good first guess.

What was now needed was a way to explain how such a diverse collection of planets reached their positions so close to their stars.

Chthonian planets

Two classes of planet had now been found orbiting exceedingly close to their stars: the hot Jupiters and hot super Earths. This gave astronomers cause to wonder if these two planetary types could be linked. Were super Earths just hot Jupiters whose giant atmospheres had somehow been siphoned away?

Evidence for this theory came from the first transiting planet to be detected: HD 209458b. In the autumn of 2003, the hot Jupiter was observed trailing its atmosphere like a gigantic comet as it crossed in front of its star. The blazing heat from its three-and-a-half-day orbit was evaporating away the gas giant's enveloping gases. If a significant amount of atmosphere were to be stripped, the planet might shrink to the size of a super Earth. The result would be either a mini Neptune or an exposed solid core. The skeletal nature of this product led the hypothetical worlds to be known as *chthonian planets*; beings of the mythological Underworld.

Despite their darkly imaginative appeal, the existence of super-evaporated chthonian planets was debatable. Hot Jupiters had such huge atmospheres that even the evaporation observed for HD 209458b might not be enough to produce a super Earth during the star's lifetime. However, evaporation was not the only way to remove an atmosphere.

As a hot Jupiter moves inwards from the outer Solar System, the pull of the star gets stronger. This causes the planet's Hill radius to shrink, so that its own gravity dominates over a smaller region. Since the planet's atmosphere has collapsed down to a size much smaller than the Hill radius, this initially makes no difference. But when the planet gets extremely close to the star, the star's gravity can dominate inside the atmosphere and begin to drag gas away from the planet. Like strong evaporation, this could leave a chthonian planet as a small gas world or an exposed core.

The fact that super Earths can be found closer to their star than hot Jupiters supports the idea of a stripped-down planet. Hot Jupiters would have their atmospheres siphoned away at 0.1–0.05au, leaving any planet that has migrated beyond that point to be a super Earth. If this proved to be true, then a rocky super Earth might be giving us a tantalising glimpse of the inside of a gas giant.

Yet there is a problem; we see very few planets in between the size of a hot Jupiter and a hot super Earth. If hot Jupiters are destined to become super Earths, we should see planets with sizes in between the two regimes as their atmospheres get stripped. Yet almost all the planets we observe close to their stars are either hot Jupiters or super Earths; just the start and end points of the chthonian theory. There is no population of hot planets between Neptune and Jupiter in size. While not completely impossible, it does seem highly unlikely that we have just missed observing these worlds. So if super Earths are not just overflowing hot Jupiters, what are the other options?

Building local

A tempting idea is to form super Earths at their current locations. If the star's natal protoplanetary disc could birth these worlds directly, it would explain why the planets are so numerous. We previously dismissed the possibility that massive hot Jupiters could form where there is so little rocky material, but was this still true for the much smaller super Earths?

Our Solar System's innermost planet is Mercury; a world just 5.5 per cent of the Earth's mass sitting at a respectable distance of 0.4au. This is more than three times further out than the main populations of both the hot Jupiters and super Earths.

At first glance, this situation ought to be the norm. The size to which a planet can grow hinges on how much material it can gather from the protoplanetary disc. This depends on the quantity of dust and planetesimals surrounding the growing planet, and the extent of the planet's gravitational reach (Hill radius). Close to the colossal tug of the Sun, the planet's gravity can control only a small region of space, restricting the amount of new material that can be reeled in for growth. Planets close to their stars should therefore be small.

But what if our Solar System was abnormal from birth? While our protoplanetary disc had very little material close to the Sun, perhaps a more typical example was rich in dust around the region where super Earths form. This would enable even a small Hill radius to gather plenty of solids.

We constructed our protoplanetary disc in Chapter 1 by taking the current positions of the planets and smearing them out around their orbits to recreate the dusty beginnings. The result was the Minimum Mass Solar Nebula. What happens if we do the same with the planetary systems than contain super Earths?

A protoplanetary disc created from crushed-up super Earths shows where the dust would need to sit to directly build that planet population. Unfortunately, piling material into a planet-making disc can lead to problems. Since a high density of dust would be suspended in a high density of gas, the inner disc now becomes very massive. Like the mechanism proposed for forming gas giant planets in the far outer disc, the inner disc can now break apart due to its own inflated gravity. Should that happen, the new planets would resemble gas giants and be completely different from the super Earths. Moreover, crushing up the planetary systems that contain super Earths produces very strangely shaped protoplanetary discs. Many of these are so weirdly proportioned that they simply could not have formed around a star at all, requiring bizarre anomalies such as the disc becoming progressively hotter further away from the star.

The conclusion is that there is no common protoplanetary disc that can birth super Earths. Instead, the mass needed for these planets must appear after the disc has formed.

The planet broom

The next idea involves a giant broom. When applied to its usual job of sweeping floors, a broom can gather dust into a pile. What seems to be very little dirt when spread evenly over the floor, can turn into a sizeable bag's worth when swept into a small area. Was there a protoplanetary equivalent of a broom that could sweep rocky grains into a pile big enough to build a super Earth?

Sweeping up rocks avoids the problems of a special super Earth protoplanetary disc. The disc could form in its regular shape, without an unstably high quantity of gas and dust close to the star. The rocky solids would then be gathered from around the disc and deposited at the orbits of the super Earths, allowing these planets to swiftly form. Since the sweep-up would involve solid particles but not gas, the inner disc would not become heavy enough to break apart into gas giant planets. The only question was what could act as a broom? The answer is a hot Jupiter.

A Jupiter-sized planet migrating towards the star will plough into the rocky planetesimals that were building terrestrial worlds. While many rocks will be scattered away or accreted into the planet, others will end up in resonant orbits with their laps around the star synchronised with the Jupiter. In this stable configuration, these planetesimals are forced inwards with the Jupiter to pile up close to the star. Now bunched together, the planetesimals collide to form a world bigger than any in our inner Solar System. The result is a super Earth orbiting closer to the star than the hot Jupiter. It seems plausible, but is there evidence that this really happens?

Gliese 876 is a star smaller and cooler than our Sun known as a red dwarf. It sits about 15 light years away in the constellation of Aquarius, the Water Carrier. Radial velocity measurements of the star's wobble have revealed four planets, with the innermost world a super Earth of nearly 7 Earth masses on an orbit lasting just two days. Slightly further out with orbits of 30 and 60 days are two hot Jupiters.

The two Jupiters and the outermost Uranus-sized fourth planet have resonant orbits. The innermost of the three giants performs four orbits in the time it takes for the middle planet to orbit twice and the outer planet to orbit once. This 1:2:4 arrangement is the same as that found in Jupiter's moons, Ganymede, Europa and Io. The resonances support the idea that the three planets migrated inwards together. Their gravitational pull on one another would have coupled their orbits (as happened to Jupiter and Saturn during their grand tack through the Solar System). As they moved inwards, smaller planetesimals could have been caught in further resonant orbits, and shovelled forwards. These rocky pieces would then collide, breaking from their resonance as they formed the inner super Earth.

Notably, the size ratio between Gliese 8/6's largest planets is very different from Jupiter and Saturn's, with the outer world being the larger at more than 2.5 Jupiter masses, and the inner planet at 0.7 Jupiter masses. This probably prevented the planets from performing the U-turn that would have stopped them from becoming hot Jupiters.

The presence of this super Earth's giant Jupiter brothers gives weight to the theory that closely orbiting planets could form from material swept up during migration. The problem is that super Earths are substantially more common than hot Jupiters. So if a Jupiter is not around to do the sweeping, how does the rocky building material reach the star?

The dead zone trap

It was the planetary system that Jack Lissauer, a space scientist at NASA's Ames Research Centre in California, described as 'the biggest thing in exoplanets since the discovery of 51 Pegasi b.'

The find was of six planets transiting the Sun-like star Kepler-11, located in the constellation of Cygnus, the Swan, around 2,000 light years away. Its announcement in 2011 hit the headlines both for the number of transiting planets in a single system and for their arrangement, which was the most tightly packed configuration ever seen.

Five of the planets around Kepler-11 circle the star within the orbit of Mercury, with the sixth only slightly outside. The tugs between the closely orbiting siblings allowed their mass to be measured via their transit timing variations. This close-knit family was revealed to have five super Earths of 2–8 Earth masses. The mass of the last planet in the system, Kepler-11g, was harder to pin down due to the weaker effect the outermost planet has on the other worlds. Estimates suggest a value of less than 25 Earth masses; a Neptune-sized world.

Here was a planetary system with not one, but six planets orbiting close to the star, and no hot Jupiter to shovel material inwards. How could such a system form? To quote Lissauer again, 'we didn't know such systems could even exist'.

While the planets of Kepler-11 were a surprise, we do know of an excellent process for shuffling rocks towards the star without a hot Jupiter: the drag from the gas headwind. As rocky boulders near 1m (3.3ft) in size, they can no longer be cradled in the flow of gas but become large enough to dictate their own orbital path. Since they do not feel the gas pressure, these small rocks move slightly faster than the surrounding gas, which results in a headwind. In Chapter 2, this was a major problem as it caused rocky material to be dragged away from where our planets sit and inwards towards the Sun. But could this process aid the production of the hot super Earths?

The main challenge with boulder drag is how to stop this flow of building material before it crashes into the star. Without the hot Jupiter to secure the rocks in resonant orbits, the rapid inflow due to the headwind would lead to incineration. What is needed is a stop sign that allows the rocks to pile up.

To form the Solar System's planets, we invoked the streaming instability, whereby pelotons of boulders gathered enough mass to shield themselves from the gas drag. However, there is no reason for the streaming instability to preferentially produce these large clusters of rocks close to the star. It is possible that rocky material might collect at the protoplanetary disc edge, beyond which the star has accreted all the gas and dust, but this would not explain a system with multiple super Earths on different orbits. Instead, a more flexible option is to use the magnetic field.

Magnetic fields are everywhere in the Universe. Take an atom, strip one of its electrons and it will have a slightly positive electric charge. Give this charged particle a push, and it will create a magnetic field. It will also feel a force from any magnetic fields already present.

On the other hand, if the atom is neutral (no electric charge) then it does not care about the magnetic field. Its movement does not create a field, nor does it feel a force within a field. This is the reason why electric and magnetic forces (known jointly as the *electromagnetic force*) in the Universe have a much smaller effect on the construction of galaxies and planets than gravity, A look at the numbers suggests that the electromagnetic force is 39 orders of magnitude larger than the gravitational force. Yet over big distances, the Universe is neutral and responds only to gravity's tug.

The heat within a star strips atoms of their electrons to create a multitude of moving charges that generate its magnetic field. These field lines weave through the surrounding gas and dust in the protoplanetary disc. The effect this has depends on the number of disc particles that are charged.

Energy radiating from the star can strip atoms in the disc of their electrons, creating charged particles of gas and dust that then become susceptible to the magnetic field. The magnetic forces ruffle up the particle orbits, aiding the accretion on to the star. Turn off the magnetic field and the rate of inward gas flow drops right down. Closest to the star, the disc feels the full force of the star's radiation. This creates plenty of charged particles to respond to the magnetic field. But before we get to about 0.1au, the star's energy will struggle to penetrate the gas all the way through to the centre of the disc. The number of charged particles drops and the gas stops feeling the magnetic field.

The region where the magnetic forces are turned off is referred to by the ominous-sounding term, *the dead zone*. Gas between the star and the dead zone edge flows easily inwards, while gas within the dead zone moves more slowly. The result is similar to a traffic jam, and the gas density rises at the dead zone edge. The gas pressure rises along with the density boost, shifting the forces felt by the gas at that point in the disc. This allows the gas to orbit at the same speed as the rock, removing the headwind on the boulders. No longer dragged towards the star, these rocks collect around the edge of the dead zone and begin to collide to birth a super Earth.

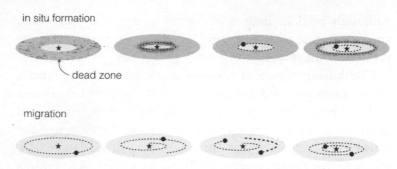

Figure 10 Hot super Earths either form from rocky material that has been swept towards the star (in-situ formation) or further out in the protoplanetary disc and migrate inwards. In one possible in-situ formation mechanism, rocky boulders are dragged inwards by the gas. These collect around the edge of the dead zone, where the disc does not have enough electric charge to feel the magnetic field. The boulders collide to form a planet, whose gravity opens a gap in the gas disc. The gap allows the star's radiation to create more charged particles, moving the dead zone outwards. Boulders can then collect at the new edge and form a second hot super Earth.

This shift in the gas flow around the growing super Earth also acts as a planet trap and throttles Type I migration. Rather that shooting off towards the (dangerously close) star, the planet can continue to grow until it can open a gap in the gas disc. This should kick-start Type II migration, but the super Earth is so massive compared with the gas this close to the star that the gas drag may be too weak to move it. Regardless of the planet's locomotion, the gap allows the radiation from the star to penetrate the disc. Dust and gas are stripped of their electrons to become charged and feel the magnetic fields. The dead zone breaks close to the planet and its edge moves outwards past the planet's gap. At this new dead zone edge, planetesimals can collect afresh to build the next super Earth. When the gas disc evaporates, a series of super Earths may be left orbiting close to the star. This appears to be very close to what we see in Kepler-11.

Although boulder drag was a promising mechanism for a production line of super Earths, the Kepler-11 system was not yet done with surprises.

Combining the mass measurements with the size from their transits revealed that none of the Kepler-11 planets were rocky. Rather, their densities suggested that they were covered in a thick gas atmosphere that envelops half the extent of the planet. The exception was the innermost planet, Kepler-11b, whose higher density pointed to a larger core filling two-thirds of the planet's size. However, even this is a much larger gas atmosphere than there is on an Earth-like world. The Kepler-11 planets were all mini Neptunes.

To match the observations of super Earths, any formation idea must be able to encompass both large rocky planets and small gas giants. So was it possible for a world born so close to its star to acquire the thick atmosphere of a mini Neptune? As it turns out, the difficulty is not how to gather gas, but how to stop.

Before the protoplanetary disc gas evaporates, a newly forming planet can pull in an atmosphere from this surrounding reservoir. On short orbits in a region packed full of planetesimals, super Earth formation should be very efficient, taking considerably less than a million years. This leaves plenty of time to accrete enough gas to become a mini Neptune. In fact, the main risk is going too far and becoming a hot Jupiter.

Hot Jupiters were previously thought to be too massive to form close to their stars. But was this assumption too hasty? With the ability to channel building material to the inner regions of the disc, would we end up with Jupiter-sized worlds?

As Jupiter grew in the outer Solar System, its gravity began to attract a large amount of gas. Eventually this became so heavy that the atmosphere went into runaway collapse, with gas piling on to the planet as the atmosphere continued to compress. This process subsided when the planet's gravity opened a gap in the gas disc. By this time, a large gas giant had been born. This looks like an unstoppable situation, but it transpires that there is a solution.

Building from piles of rocks that have been swept inwards to the dead zone, a young super Earth's atmosphere is heavy with dust. This prevents the planet's gaseous envelope from cooling efficiently, since the dust grains block radiation escaping the planet (in more technical language, the atmosphere has a high *opacity*). The higher temperature of the gas supports it against the planet's gravity, delaying runaway collapse until after the gas disc has evaporated. The result is that the planet is able to gain a thick atmosphere, but nothing close to the drowning gases of a hot Jupiter.

Whether a super Earth becomes a giant terrestrial planet or a small gas world could depend on the protoplanetary disc. Heavier discs can assemble a super Earth planet more rapidly, allowing more time to grasp bigger atmospheres. For lighter discs, the super Earths might not form until the gas was close to evaporation, leaving these planets rocky, with thinner atmospheres.

Building planets in their observed location is known as *in-situ* formation. The hot Jupiter shovel and boulder drag make *in-situ* formation for super Earths a serious possibility. But this did not completely seal the deal for super Earths.

While Kepler-11 became the prototype for stars closely orbited by tightly grouped planets, it was far from alone. A year after its discovery, Kepler-32 was found, with five planets all smaller than 3 Earth radii and with orbital times of 0.7–22 days. Then three more planets were found surrounding the star HD 40307, bringing its total up to six planets with less than about 7 Earth masses, five of which have periods of 4–52 days. Other systems followed, suggesting that more than 10 per cent of stars might have a similar configuration of worlds.

So if this was a common outcome for planet organisations, why again was our Solar System different? Our gas giants may have avoided becoming hot Jupiters, but the early days of the Solar System should have seen a stream of planetesimals flowing towards the star. Yet we failed to form even one super Earth.

There was also debate about whether a planet forming *in-situ* could hold on to the large atmosphere of a mini Neptune. With a formation site packed with material, but the planet's Hill radius still small, a collection of Earth-sized embryos might initially be produced (due to an Earth-sized isolation mass, in the language of Chapter 2). This would be followed by an extended period of giant impacts between the embryos to form the super Earth. These large collisions risk vaporising the gases of the new world, leaving only a thin envelope surrounding a rocky planet.

Possible explanations existed to weave around these problems: the dead zone may change between protoplanetary discs, another process might interfere with the planetesimal flow, and giant impacts may not always be necessary. But this did provide enough doubt to explore other options.

A migrating population

Away from any planet traps, the chunky mass of the super Earths should result in swift migration. Did this make it likely that the super Earths migrated towards the star from much further out?

The idea that super Earths were born far away from the star has nice and not-so-nice features. The lack of a clear divide in the size of rocky super Earths and mini Neptunes suggests that these planets are a single class, forming through the same mechanism. Since our own Neptune formed past the ice line in the outer Solar System, it seems reasonable that hot mini Neptunes, and therefore the rocky super Earths, might all begin in a similar location. The rocky planets would be the versions that failed to acquire huge atmospheres, due either to lack of mass or from having formed close to when the gas disc was evaporated.

This also ties their evolution to the hot Jupiters. In all cases, large planets could start beyond the ice line, where there is plenty of planet-building material. Far from the star, the planet's Hill radius is large and it can swiftly gather mass and gases, and avoid atmosphere-stripping collisions.

This also provides an explanation for our Solar System's lack of super Earths. Jupiter and Saturn's U-turn grand tack prevented the migration of Uranus and Neptune. Without our largest gas giants barring their path, the embryos of these smaller worlds might have travelled inwards towards the Sun. In one movement, we have removed the need for a slew of different planet-formation mechanisms to explain these varying worlds.

Yet this assertion about the universal power of migration is a bold statement. It would mean that major planet reorganisation is a common occurrence. While hot Jupiters most probably migrated inwards, they only appear around 1 per cent of stars. On the other hand, hot super Earths are thought to orbit around 50 per cent of stars. For all these planets to have changed orbits, migration must not only be possible, but it must be a major player in planetary-system architecture.

There are also some observations that do not match this picture. As with Jupiter and Saturn's brief spate of migration, the tugs between neighbouring planets moving through the gas disc should result in resonant orbits. The outer planet should approach its inner neighbour either during migration, or when the innermost world stops at the protoplanetary disc inner edge. Their orbits will then slide into resonance, with an exact integer ratio between their orbital times. While this is seen for a few systems, such as Gliese 876, many others, such as Kepler-11 and HD 40307, show no such pattern. So does this mean that migration definitely did not occur?

Although resonant orbits would support the migration picture, it turns out not to be a deal breaker. One reason for this is that Type I migration is an infuriatingly picky process. As we have seen with planet traps, the first stage of migration is very sensitive to the conditions of the surrounding gas. It also depends on planet mass. The more massive the planet, the stronger the gravitational pull with the gas disc. Heavier planets typically migrate faster, until they are able to open a gap in the gas and slow to Type II migration. However, there can also be regions in the disc where a combination of the planet's pull and local gas conditions can flip the direction of migration for a short period. This leads to migration paths depending strongly on the specific combination of forces from the planet's current mass, gas conditions at its present location and any pulls from neighbouring planets.[*]

Such tailor-made migration tracks allow more flexibility in the positions of planets when the gas finally evaporates. In one set of computer models for this contrary process, a region for reverse migration developed for planets above 5 Earth masses. The planetary embryos able to grow fast enough to hit this sweet spot were potential hot Jupiters. These most massive worlds were then delayed in reaching the star to leave them stranded slightly further out than the smaller super Earths. This matches up with the observation that hot Jupiters cluster behind the super Earth population. It also supports their relative rarity, since rapid growth was needed to hit the region of outward migration with enough beef to be turned around. Such varying paths are also more difficult to lock into resonance, leaving planets with a broad range of separations.

Other methods joined this idea in support of super Earth migration. Another possibility was that the planets could initially have been in resonance, but were later knocked about by bombardment from the remaining rocks once the gas had evaporated. This follows the evolution of our own giant planets, whose orbits shifted as they scattered planetesimals. Alternatively, the closely orbiting super Earths could be affected by an unseen giant planet further from the star. In a distant location more difficult to detect, the presence of a big gravitational bully could ruffle the orbits of super Earths to break up their resonances.

With the idea of migrating super Earths firmly on the table, a second concern arose. If migration was a major way to

---

[*] The sensitivity of Type I migration to the exact conditions around the planet makes it a big can of worms.

form super Earths, could any system with a closely orbiting planet support a habitable world like our own Earth?

☄

Did Saturn save our planet's bacon? Without the second gas giant, Jupiter would have followed the fate of worlds such as 51 Pegasi b, crashing through the inner Solar System on the way to the Sun. It may well have been followed by Uranus and Neptune, migrating to become closely orbiting super Earths. As they travelled inwards, these giant worlds might have battered our precious Earth to pieces.

The Earth's location at 1au is key to its habitability. This distance from the Sun means that the planet receives just enough heat to be neither too hot nor too cold to support our existence. If it could not form at this position due to the migration of outer planets, there is a high chance life would never have evolved.

So can a world like our Earth exist behind closely orbiting hot Jupiters or super Earths? If not, we may be forced to dismiss half of all planetary systems in a search for alien neighbours. That might make life in the Universe rare indeed.

The inward sweep of a planet is potentially a disastrous event. The gravitational pull from a migrating world will scatter rocky material out of the inner planetary system, shovel planetesimals towards the star and devour a sizeable chunk of the rest. The terrestrial planet zone would be left an empty factory, barren of planet-building material.

Should a young planet have formed before the migration, the sudden pull of an approaching big planet would scatter it on to a new orbit like a comet. A scattered path around the star risks being strongly elliptical, with the planet's distance from the star varying strongly during its circuit. The result can be extreme seasons as the surface temperature soars and plummets over a planet's year. It may not be impossible to retain water and develop life in such a condition, but it is going to be difficult.

The prospect is bleak but there is still a glimmer of hope. If enough dust and rocks are left behind the migrating planet, then the building of terrestrial worlds can begin afresh. How much material is left for a restart will depend on how fast a migrating planet sweeps through the system. The precarious Type I migration rates make this hard to estimate, but a planet that lingers around the terrestrial planet-making zone will scatter away more rocky material than a migrating world that zips more quickly towards the star.

Rocky planetesimals that are scattered may also be able to return to more circular orbits due to the gas disc. A planetesimal on an elliptical path is forced to cut across the circular gas flow of the disc. The different speed of gas and solid creates a very strong drag, drawing the rocks back on to circular paths to continue the planet-forming process.

All is also not entirely lost for a scattered planet, which may yet be able to recover a more circular orbit. On a bent elliptical orbit, the planet feels a varying pull from the star as it approaches and then moves further away. As in the case of the hot Jupiters scattered inwards by the Kozai–Lidov mechanism, this fluctuating pull can circularise the planet's orbit once again. The gas disc will also resist the planet's orbit becoming elliptical, helping it to maintain a circular path.

Recovery after a migrating planet population may even hold some advantages. A second generation of planets may not reach Mars size until after the evaporation of the gas, removing the need for planet traps to prevent migration towards the star. The large scattering of rocks from the first migrating planets may also deliver ices to the inner system, allowing water-rich worlds to form. The result would be hot planets close to the star, but worlds with more promising conditions (albeit difficult pasts) for life behind them.

An unresolved mystery

The formation of the hot super Earth population remains an intriguing problem. Did they migrate from behind the ice line, or form from planetesimals and boulders shovelled inwards by hot Jupiters or gas drag forces?

One way to untangle these ideas is to hunt for the fainter signatures of outer planets. A star with planets both far out and on close orbits is less likely to have seen strong migration than a star surrounded only by close-in planets. A planet that has formed far from the star will also be heavy in ices. This may produce an atmosphere thick with water vapour that could be glimpsed by the next generation of telescopes. However, until we can unravel the meandering ways of both planets and planetesimals, our most common planet class will remain a mystery.

CHAPTER SEVEN

Water, Diamonds or Lava? The Planet Recipe Nobody Knew

Two years after the discovery of 51 Pegasi b was announced, astronomers were becoming skilled at picking out the wobble in a star's position that hinted at the presence of a planet. The result was six more exoplanet finds, all hot Jupiters like 51 Pegasi b, whose huge mass and close orbits made them the easiest to detect.

The last three of these were announced together in 1997. Among the trio was HR 3522b, a planet slightly smaller than Jupiter on a 14-day orbit. Although the planet was referenced in the journal paper by its listing in the Yale Bright Star Catalogue (HR), it would become more commonly referred to as 55 Cancri b; the first planet found around the 55th star in the constellation of Cancer, the Crab. Its discovery was instantly notable as one of the first planets found outside our Solar System, but this would not be the only record for the planetary system. In fact, 55 Cancri b was the first planet found in a system more alien than anything we could imagine.

Within 10 years of the discovery of 55 Cancri b, four more planets had been detected orbiting the same star. This made 55 Cancri the first star to be found hosting five planets, and also one of the first three to have a super Earth with a mass similar to Neptune; 55 Cancri e is about 8 Earth masses (48 per cent of Neptune's mass) with an incredibly short orbital time of only 18 hours, sitting at just 5 per cent of Mercury's distance from the Sun.

Observations of 55 Cancri had revealed it to be one half of a binary. Its stellar sibling is a smaller red dwarf star that sits at a distance of more than 1,000au. While too small and distant to damage the planets forming around its bigger sibling, the regular pull from the red dwarf seems to be causing the planetary system to turn on its head.

This is a similar effect to the Kozai–Lidov mechanism that we met in Chapter 5 as a way of using a stellar companion to move hot Jupiters towards their stars. For the planets of 55 Cancri, the gravitational pulls from the neighbouring worlds hold the planets' orbits together, so the entire system slowly flips over like a set of synchronised swimmers. If it were possible to stand on the surfaces of these planets and stare at the sky, the constellations would appear to slowly shift as the planetary system performed a somersault. Admittedly, it would take a particularly long-lived life form to notice this, as a full flip would take around 30 million years.

Yet what made the system truly strange was the properties of its super Earth, 55 Cancri e. Like the other planets in the system, 55 Cancri e had been found via the star's induced wobble with the radial velocity technique. In 2011, NASA's Spitzer Space Telescope also observed the super Earth transiting across the star's surface. This was another 'first' for the planetary system, since 55 Cancri is visible with the naked eye, making its innermost planet the first to transit a star that can be seen without a telescope. The transit observation measured the planet's radius and the angle of the orbit, pinning down the planet mass. The super Earth's size was discovered to be 20 per cent larger than the Earth at 2.2 Earth radii. It was another of those values that would appear to make absolutely no sense.

With both mass and radius measurements, the density of 55 Cancri e could be easily calculated as $4g/cm^3$. Despite a radius that suggested the planet should be a mini Neptune,[*] this density was far too high for a gaseous world with a large hydrogen and helium atmosphere. An 8 Earth-mass mini Neptune should have a density of only $1.3g/cm^3$. On the other hand, $4g/cm^3$ was far too low for a rocky Earth-like

---

[*] In Chapter 6, we said a rough rule of thumb was that above a size of about 1.5 Earth radii, the planet was probably gaseous rather than rocky.

interior in a planet of that mass. While the Earth's density is moderately close at $5.5g/cm^3$, a planet eight times heavier would compress the rock to above $8.5g/cm^3$. This meant that 55 Cancri e was too small for a gas world and too large for a rocky world. So, what was it?

If this super Earth was a rocky planet, then the ingredients for its composition should match our terrestrial worlds. This indicates a bulk made predominantly from iron and silicates. For a given mass, the smallest rocky planet possible would be one made of pure iron. The largest would have no iron at all, but be made throughout from the lighter silicate rocks. Neither of these extreme versions is very likely. The elements of our terrestrial worlds condense into solids at a sufficiently similar temperature to create a mixed planet-forming flour. Our hottest and most iron-rich planet, Mercury, still retains 30 per cent of its mass in a silicate mantle. Yet even these most improbably extreme examples remain smaller than 55 Cancri e.

What if we attempted to reconcile the radius and mass of 55 Cancri e by abandoning the distinction between terrestrial and gaseous planets and forming a hybrid world? Such a planet would have a rocky core substantially bigger than the Earth's and be capable of retaining a primitive thick hydrogen and helium atmosphere. Since the gases are so light, even an atmosphere containing just 0.1 per cent of the planet mass could result in a world with the density matching 55 Cancri e.

The hybrid planet would be perfect if it were not for 55 Cancri e's incredibly short 18-hour orbit. With a year lasting less than an Earth day, the super Earth is looping its star at a distance of just 0.016au. This proximity to a burning nuclear engine gives the planet an estimated average temperature of around 2,000°C (3,600°F). At these blistering values, the planet's mass will struggle to stop a light atmosphere of hydrogen and helium evaporating away. Such an atmosphere should be burned off within a few million years, which is short in planet-formation terms. This makes it unlikely that we would catch sight of 55 Cancri e while it still had a primitive envelope of gases.

☄

With a Neptune-like atmosphere out of consideration, what could be less dense than rock but still heavy enough to be held by the planet? The answer might be a very strange state of water.

If 55 Cancri e had originally formed beyond the ice line, then it would have been born packed full of ice. As it migrated towards the star, the hydrogen and helium in its atmosphere would be lost to leave a rocky core under a water-vapour envelope thousands of kilometres thick. The idea that a planet orbiting so close to the inferno of its star could be covered in water is indeed bizarre. The water on this planet would certainly not resemble the cool liquid that comes out of a kitchen tap. Instead, the water on 55 Cancri e would exist in an extremely rare state known as *supercritical*.

Supercritical fluids occur at very high temperatures and pressures. For example, rocket fuel is in a supercritical phase when it is blasted from the tail of a launching spacecraft. In this form, the distinctions between liquids and gases become blurred to leave a substance somewhere in between these definitions. On such a world, it would be impossible to tell where the oceans met the sky. If it were possible to survive, you would find yourself suspended somewhere in the supercritical fog.

A burning-hot world with an 18-hour year, enveloped in a liquid-like gas on an orbit that slowly flips over, is a strange place indeed. Yet, there is an even weirder explanation for the composition of 55 Cancri e. Rather than supercritical water, the super Earth could be heavy in diamonds.

☄

While 55 Cancri is a star similar in size to our Sun, its composition is not identical. Rather, 55 Cancri is suspected of being rich in carbon. Until they reach old age, stars mainly consist of hydrogen and helium with trace amounts of other elements, such as carbon, oxygen, magnesium, silicon and iron. The Sun has roughly half as much carbon as oxygen, a trait that is typically expressed as the ratio between these two elements, $C/O = 0.5$. In contrast, observations of 55 Cancri in 2010 suggested that the star has actually slightly more carbon than oxygen, with a ratio $C/O = 1.12$. These differences in stars are important since the protoplanetary disc is made from the same material. Therefore, if the star is carbon rich, it is likely that the grains of dust that form the planets will also be enhanced with carbon. This could build worlds out of very different material from our terrestrial population.

Despite its importance for biological life, the Earth is surprisingly carbon poor. Ninety per cent of the Earth's mass is in iron, silicon, oxygen and magnesium. The majority of the iron is in the Earth's core, with the remaining elements forming the silicate mantle and crust. Carbon is only a minor constituent, making up less than 0.2 per cent of the Earth's mass. This tiny fraction is because carbon only condensed into solid grains in the cold outer Solar System. Around the forming terrestrial worlds, it remained a vapour and was blown away as the Sun dispelled the gas disc. As is the case with the possible origin for our oceans discussed in Chapter 4, the Earth began devoid of carbon and received its meagre quantity from meteorites sailing inwards from the outer Solar System.

Should the fraction of carbon get a boost and rival (or even exceed) the fraction of oxygen atoms in the protoplanetary disc, the solid building material begins to change. With carbon atoms controlling the scene, silicon starts to bond with carbon rather than oxygen to form solid silicon–carbide instead of silicate. Planets born from this dust would be made from carbon and the new silicon–carbide, rather than oxygen compounds.

If 55 Cancri e has such a carbon–rich interior, the need for a volume-boosting envelope of lighter material vanishes. For the planet's observed mass, a solid body made from iron, carbon and silicon could produce the correct radius. Not only does this remove the need for supercritical water, but 55 Cancri e might not have any water at all.

In a carbon-rich protoplanetary disc, the oxygen would be grabbed by the carbon to form the toxic carbon monoxide. There would be little oxygen left to bond with hydrogen and form water. Even the outer planetary system might therefore have no water ice. Modelling the planetesimals forming in carbon-rich systems, Torrence Johnson of NASA's Jet Propulsion Laboratory in Pasadena, California jibbed, 'There's no snow beyond the snow line.'

The lack of water in the planetary system means that even if 55 Cancri e were on an orbit that allowed a more agreeable climate, due to its carbon base it might be incapable of supporting any life that we currently recognise. Collaborating with Johnson, Jonathan Lunine of Cornell University commented humorously, 'It's ironic that if carbon, the main element of life, becomes too abundant, it will steal away the oxygen that would have made water, the solvent essential to life as we know it.'

Aside from having no water, what would a carbon world be like? The crust would probably be graphite, the substance that is found in pencils. Beneath the planet's surface, the pressure would mount to produce a mantle of diamond. Much of the carbon in the Earth's mantle also exists as diamond, oxidising to carbonate near the crust as the pressure drops. The reason why we are not rolling in gems is that the total quantity of carbon is so low, below 0.2 per cent compared with just over 50 per cent oxygen. On a carbon world, the quantity of diamond would cause these glittering gems to erupt from the ground during volcanic activity.

If liquid existed on the planet, then it would also be a carbon-based substance such as a sea of tar. The atmosphere would be heavy in carbon monoxide and dioxide, while carbon-heavy rain would make the air thick with smog. This is the optimistic picture. The planet may actually have no atmosphere at all.

The interior of the Earth is an active machine. The crust is split into rigid chunks known as tectonic plates. Beneath these is the planet's mantle. While the Earth might appear solid, over geological timescales of millions of years, the mantle moves as a highly sluggish fluid. This acts as a conveyer belt for the tectonic plates and shuffles them around. When two plates move apart, the underlying mantle is exposed and cools to form fresh crust. Older and thicker crust is melted where plates slide under one another, often creating volcanoes at this boundary. These movements of the crust and mantle circulate the planet's atmosphere and nutrients, and even generate the magnetic field. However, replace our mantle with diamond, and this essential motion becomes much more difficult.

Diamond has a very high viscosity; a fluid friction that controls how sluggishly materials flow. Syrup is more viscous than water, and a diamond mantle is about five times more viscous than a silicate layer. If the carbon fraction of a planet exceeds 3 per cent, the difficulty in moving the mantle can draw the moving plate tectonics to a screeching halt.

The absence of plate tectonics would give a planet an immobile lid, making it far harder for volcanoes to form. While a decrease in the number of explosive mountains might seem like a positive attribute, it also suppresses the development of an atmosphere. The resulting world would be a stagnant body, heavy in gems but lacking in air.

The graphite crust of the planet might also make it too hot. Even on an Earth-like orbit, the dark colour of the graphite would absorb, rather than reflect, sunlight. The planet would be the black car in a Florida car park,* compared with the green-and-blue Earth version. This would be a problem for maintaining surface liquid water, even if it were possible to somehow acquire oceans.

* The rush to park under the one tree in a Florida supermarket car park is real, no matter how far it is to the store's entrance.

All of this considered, the lure of a planet of beautiful gems is not sufficient to make a carbon world an appealing property.

🪐

Before 55 Cancri e could officially be declared a carbonic hellhole, its carbon credentials were called into question. The problem was that measuring a star's C/O (carbon to oxygen) ratio is a difficult business. To point the finger of blame more directly, it is measuring the star's oxygen content that is particularly tricky.

Stars consist of incredibly hot, dense cores surrounded by a slightly cooler atmosphere of low-density gas. In the Sun, the temperature of the core can reach more than 15,000,000°C (27,000,000°F), while the outer layer is at 5,500°C (10,000°F). This outer atmosphere is known as the *photosphere* of the star. The photosphere's temperature is still intense, but it is cool enough to allow atoms to hold on to their electrons. These electrons are arranged on a ladder of unequally spaced energy levels. As the radiation pours from the core, the atoms absorb wavelengths with the right energy to flip an outer electron on to one of the higher energy rungs. Which wavelengths are absorbed depends on the energy levels occupied by electrons, and therefore the type of atom. By examining a star's light and seeing which wavelengths are absent, the different atoms in the star can be revealed.

Difficulties arise when two different atoms absorb at very similar wavelengths. It can become hard to separate out the two, creating uncertainties in the quantity of each atom. In the case of the C/O measurement for 55 Cancri, the commonly measured wavelength for oxygen was extremely close to one for nickel. In 2013, data from the star was freshly examined. Rather than relying on differentiating between the main oxygen and nickel wavelengths, three different wavelengths that oxygen atoms can absorb were compared to the absorption from nickel. The conclusion was that the C/O ratio in the star was much lower than the initial 1.12, at around 0.78. Carbon replaces oxygen in silicon compounds about when the protoplanetary disc gas has C/O ~ 0.8. This new value therefore put 55 Cancri e right on the brink of being a carbon world, with the answer resting on a particularly challenging observation.

For those desiring an abhorrent carbon world, a further complication could save the situation. While the star and the protoplanetary disc are born with the same composition of atoms, the solid particles in the disc change over time.

When exploring the formation of our protoplanetary disc in Chapter 1, we noted that the material of the dust grains depended on temperature. Closer to the Sun, silicate and iron compounds that only vaporise at high temperatures are present. Volatile molecules such as water remain as a gas until the temperature drops past the ice line. Yet this transition from gas to solid does not happen instantly. The ages of meteorites that have fallen to Earth suggest that contrary to appearing in a single burst, the solid particles that built our planets condensed out of the gas over 2.5 million years. This is long enough for conditions in the protoplanetary disc to change to give carbon-heavy planetesimals a boost.

When the C/O ratio sits below 0.8, carbon remains a gas through much of the protoplanetary disc. Silicon grabs the oxygen to form grains of silicate, leaving the carbon untouched. This steadily depletes the gas of oxygen and the C/O ratio begins to rise. Solid particles forming later in the disc may therefore come from a gas so carbon heavy that graphite and silicon carbide preferentially form.

This means that even if the protoplanetary disc gas initially has a C/O ratio of less than the magic 0.8 value, it may still end up with a large quantity of solid carbon. Calculations suggest that a value as low as C/O = 0.65 is sufficient to produce carbon-rich planetesimals. Not only could this make 55 Cancri-e a carbon world, but it would not be alone.

The C/O ratio in our local stellar neighbourhood suggests that a third of planet-hosting stars might have a C/O ratio above 0.8 and harbour dastardly carbon worlds. Even if this value proves to be an over-estimate due to the difficult oxygen measurement, carbon planets may still form a significant population. Two of the stars found with notably high C/O ratios are hosts to gas giants. HD 189733 is 63 light years away in the constellation of Vulpecula, the Fox. Its gas giant is a hot Jupiter with an orbit lasting 2.2 days. HD 108874 is 200 light years away in the constellation of Coma Berenices (Hair of Berenice, an Egyptian queen), and has two Jupiter-sized worlds that sit further from the star. HD 108874b sits at the Earth's location of 1au, while the second planet orbits at 2.68au. As gaseous worlds, none of these planets would have a solid surface. However, should they harbour moons, these satellites may well be more carbon worlds.

Rocky recipes

Should 55 Cancri e avoid the fate of a carbon world, silicon would combine with oxygen to form silicate rocks. This initially sounds far more Earth-like. However, it turns out that not all rock recipes are created equally.

Silicates are formed from silicon and oxygen combined with another element. In the Earth's mantle, that addition is commonly magnesium (although sometimes iron). The exact blend of mineral depends on the relative abundance of magnesium and silicon, expressed as the ratio Mg/Si. The Sun has an Mg/Si ratio of a little over 1, giving roughly equal numbers of both atoms. This produces a mix of pyroxene (one silicon atom and one magnesium per molecule) and olivine (two magnesium atoms to one silicon) through our mantle. Surveys of magnesium and silicon in neighbourhood stars suggest that the Mg/Si ratio is highly variable. This could result in terrestrial worlds made from silicate rock, but with different mineral versions for their interiors.

For an Mg/Si ratio of less than 1, the silicon fraction tops the magnesium. Silicon mops up the available magnesium to form the familiar pyroxene, but the remaining silicon bonds with other minor elements such a potassium, aluminium, sodium and calcium to create a family of minerals known as feldspars. Feldspar is a common mineral in the Earth's crust, so a magnesium-poor planet might have a mantle made of crust material. On the other hand, if magnesium exceeds silicon then the magnesium-rich olivine and ferropericlase (magnesium and oxygen) minerals are produced.

The reason why all these mineral types matter comes back to the ability of the mantle to move around. A mantle more or less viscous than the Earth's minerals risks changes to the crustal plate tectonics and volcanic activity to produce entirely different surface conditions. Messing with the mantle recipe is like attempting to make pastry from flour and butter; get the ratios wrong and you may end up with something ill suited to support life.

If 55 Cancri e forms silicates, then its fate may be a sluggish mantle. The star has a measured Mg/Si ratio of just 0.87, making any silicate material magnesium poor. The crust-like composition of feldspars would give the mantle a thick viscous flow compared with the Earth's rock. This may result in explosive volcanism, as gases are unable to escape the slow gelatinous magma, carrying the molten rock upwards during the eruption.

On the other end of this scale are the magnesium-rich planets orbiting the star Tau Ceti. The system sits 11.9 light years from Earth in the constellation of Cetus, the Whale. As the closest star to the Sun after Proxima and Alpha Centauri, the planets of Tau Ceti have long been ammunition for science fiction.

The existence of the planets is controversial, but there is evidence that five super Earths orbit Tau Ceti. The inner three planets sit close to the star, but the outer two potentially have surface temperatures comparable to those of the Earth. However, Tau Ceti has a Mg/Si ratio of 1.78, making it 70 per cent more magnesium rich than the Sun. If these worlds are rocky, they will harbour magnesium-heavy mantles of olivine and ferropericlase. Unlike feldspars, ferropericlases flow more easily than the Earth's mantle mix. This could give a more vigorous stirring of the planet's interior, affecting the movement of the tectonic plates. Alternatively, the magnesium-rich crust may be thick and unable to fracture into tectonic plates at all. If this still produces volcanic activity, the gas will easily escape the freely flowing magma to spawn non-explosive effusive eruptions of lava.

Although we do not yet understand exactly how these different geologies will change a planet, there is a bottom line: an Earth-sized rocky world does not equal an Earth.

Explosive discoveries

The nature of 55 Cancri e continued to niggle at astronomers. Was this a carbon world with a diamond mantle, or a planet with a silicon-rich surface coated with exotic seas of neither liquid nor gas?

The possibilities took an entirely new twist due to fresh observations published by a team from the University of Cambridge in 2016. Training the Spitzer Space Telescope on 55 Cancri, the astronomers examined the light not just as 55 Cancri e transited the star's surface, but also as it dipped behind the star. Both overlaps between the star and planet result in a drop in the observed brightness. When the planet is neither in front of the star nor behind it, the light observed is a combination of the star and the planet's own dim radiation. During a transit, the planet conceals part of the star's surface and the brightness drops. When the reverse happens and the star conceals the planet, a second smaller dip is also seen as the planet's radiation is obscured. This second transit is known as an *occultation*, or *secondary eclipse*.

The light from the planet is from the reflected starlight and the planet's own thermal radiation. As the planet is colder than the star, the thermal part of its radiation is in the longer, infrared wavelengths that the Spitzer Space Telescope is primed to see. By observing the change in this thermal radiation as the planet disappeared from sight during the occultation, the planet's temperature could be measured.

Observations of the occultations of 55 Cancri e between 2012 and 2013 revealed something decidedly odd; the thermal radiation from the planet was fluctuating by 300 per cent. The resulting calculated temperature for the planet varied between 1,000–2,700°C (1,800–4,900°F); a massive change of nearly 2,000°C (3,600°F). In addition, the dip in brightness during the planet's transit in front of the star was also not constant, showing a variation in how much light the planet's surface seemed able to obscure as if it were changing in size.

These massive fluctuations of incredibly high temperatures suggested a new idea: volcanoes. Bathed in this heat, almost any type of rock would melt. The result would not be a planet with oceans of strange water or tar, but one swimming in magma. With a molten crust, eruptions can violently spew plumes of melted rock up into the atmosphere.

Such activity is not alien to our own Solar System. Jupiter's third-largest moon, Io, is the most volcanically active world in our planetary system. Volcanoes on the moon spew material more than 300km (190mi) above the surface. So far from the Sun, the energy to melt Io's lower crust and drive the volcanoes stems from the tidal heating due to the moon's slightly elliptical orbit.[*] The varying gravitational tugs come from Jupiter and the two neighbouring moons, Europa and Ganymede, which flex Io's surface and cause it to distort by up to 100m (328ft). This is more than five times larger than the difference in the height of the ocean tides on Earth. On 55 Cancri e, the pull from the star and the neighbouring planets could create a similarly elliptical orbit that flexes the planet and drives some serious volcanic action, in addition to the star's melting heat.

If the volcanic ejecta were thrown high into the atmosphere of 55 Cancri e, the lower layers of the atmosphere would be obscured. Observations would then detect the higher, cooler parts of the planet's gases and cooling plume of ejecta to provide a lower estimate of the planet's temperature. As the volcanic activity eased, the plumes would disperse and the planet's lower and hotter atmosphere would once again

---

[*] We met tidal heating for the puffy hot Jupiter, WASP-17b, in Chapter 5. Due to the planet's elliptical orbit, the gravitational tug from the star varied in strength, flexing and heating the planet. This is discussed in more detail later.

become visible. The measured temperature fluctuation could therefore be due to bursts of volcanic activity. These plumes could also explain the change in the light dip as the planet transits across the star. When the air is thick with volcanic ash, the star's light is blocked more effectively to give a larger radius estimation for the planet.

To match the observed changes in both temperature and radius, the volcanoes on 55 Cancri e would have to throw material higher than any volcano in the Solar System. For plumes throwing up all over the planet, the average height would need to be of the order 1,300–5,000km (800–3,100mi). If a single massive plume had to do the job, then it would need to be a staggering 10,000–22,000km (6,200–13,600mi); one to two times the radius of the planet. For comparison, Io's 300–500km (190–300mi) plumes are between 16 and 27 per cent of the moon's radius. Yet 55 Cancri e is clearly a land of extremes, so perhaps we should expect nothing less.

Should the volcanoes be true, they also cast a light on the planet's composition. Taking the actual radius of the planet to be the smallest measurement (corresponding to a plume-free atmosphere), the planet's density becomes consistent with an Earth-like iron and silicate interior, although the planet would still poorly resemble our own, with a surface of melted rock creating a world of molten lava. That said, the volcanic model does not rule out the carbon world or supercritical water envelope. To ascertain that, the content of the volcanic outflows or planet atmosphere would need to be probed.

Nikku Madhusudhan, a member of the Cambridge University team and lead scientist on the suggestion of 55 Cancri e as a carbon world said, 'That's the fun in science – clues can come from unexpected quarters. The present observations open a new chapter in our ability to study the conditions on rocky exoplanets using current and upcoming large telescopes.'

Lava worlds

It is a difficult call, but the prospect of 55 Cancri e being a volcanic ocean of lava might be even more hellish than that of a smog-filled carbon world. Yet it was not the first planet to be suspected of possessing an inferno-riddled landscape.

In February 2009, a new planet was reported transiting a star 489 light years away in the constellation of Monoceros, the Unicorn. The planet was denoted CoRoT-7b, named after the French space telescope that made the discovery: the *COnvection ROtation et Transits planétaires* observatory, or in English, *COvection, ROtation and planetary Transits*. CoRoT was designed to find transiting exoplanets on short orbital paths lasting less than 50 days; an objective that was satisfyingly fulfilled with the discovery of CoRoT-7b. The planet orbited its star in 20 hours, at just 4 per cent of the distance between Mercury and the Sun.[*]

Unlike in the case of 55 Cancri e, the density of CoRoT-7b seemed more easily explainable. The same year as its transit detection, radial velocity measurements estimated the planet's mass. CoRoT-7b has a mass just under 5 Earth masses and a radius of 1.7 Earth radii to give a density around $6g/cm^3$. This is a little light for an Earth-like iron core and silicate mantle in a 5 Earth-mass world, but possible if the iron were depleted or the observed radius a slight overestimate. It was certainly too high a density for a gaseous planet, and the measurement made CoRoT-7b the first rocky planet to be confirmed outside our Solar System.

The close density match with the Earth spawned a number of news reports (including – it must be admitted – from NASA), claiming that the new world was the 'most Earth-like planet' discovered at the time. While perhaps this was true with regard to its composition, the fact that the planet's average surface temperature was around 2,000°C (3,600°F) really made the comparison rather poor. If it were possible to stand on the surface of the planet, the star would appear more than 300 times larger than the Sun looks in our own sky. However, the standing part would be difficult because the

---

[*] CoRoT-7b is a little smaller than the Sun, but larger than 55 Cancri e. This causes its nearest planet to sit slightly closer than 55 Cancri e, but still have a slightly longer orbital time.

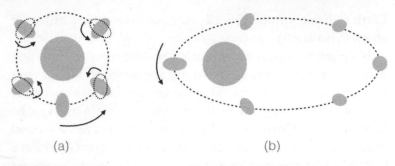

(a)   (b)

Figure 11 Tidal forces: (a) A closely orbiting planet (or moon) will be distorted by the gravitational pull of the star. The star's gravity pulls on the raised bulge so the same side of the planet always faces inwards in a tidal lock. (b) Planets on elliptical orbits are flexed as the gravitational pull from the star changes with distance. This causes tidal heating of the planet. (Actual distortion exaggerated.)

temperature would have melted a rocky surface to an ocean of lava. CoRoT-7b was not just the first confirmed rocky world; it was also the first lava world.

Lava worlds such as CoRoT-7b and 55 Cancri e sit so close to their stars that they become *tidally locked*. Planets in this configuration have one side of their surface always facing their star. Our own Moon is tidally locked to the Earth, so we only ever see half its surface. The first time a human laid eyes on the reverse side of the Moon was during the *Apollo 8* mission when the famous *Earthrise* photograph was captured.

Tidal locking occurs as the planet (or moon) gets distorted by the star's (or planet's) gravity. A spherical planet will be pulled into a slightly oblate (rugby-ball) shape by the strong tug of the very close star. As it attempts to rotate, the star's gravity will drag on the bulge raised on the planet's near side. This forces the planet to rotate so that the bulge is always facing the star. The result is a tidal lock.

The gravity of the planet (or moon) can also distort the star (or planet). In the case of CoRoT-7b, due to the huge size difference between the planet and star this reverse tug goes unnoticed. On the other hand, the Moon's pull on the Earth causes both the land and the seas to rise and fall as tides. The

Earth's distortion of the Moon raises a surface bulge of about 0.5m (1.6ft).

If a planet's orbit is not circular, then the distortion the star induces changes during the orbit as the gravitational pull weakens and strengthens. This causes the flexing that drives the volcanic activity on Io and 55 Cancri e. It may even be happening to CoRoT-7b, due to the presence of a second planet on a more distant orbit. This bigger sibling could force CoRoT-7b's orbit into an ellipse, causing the planet to flex as it moves further and closer to the star. The surface of CoRoT-7b might then match 55 Cancri e as a volcanic nightmare.

The further apart the star and planet orbit, the weaker this tidal distortion. The drag on the raised bulge is then too weak to create a lock, which is why the Earth is not tidally locked to the Sun. If the two orbiting objects are similar in size, they can be tidally locked to one another. This is the case for Pluto and its large moon, Charon, which face one another like partners in a rumba.

Planets tidally locked to their stars are split worlds, with one side of perpetual day and the other of eternal night. How well heat is distributed between the day and night sides depends on the planet and the ability of its atmosphere to circulate the warmth. Due to the vast distance of CoRoT-7b from us, we are not able to detect its atmosphere or any variations in its temperature. If the atmosphere were able to even out the heat, the surface would be expected to reach about 1,500°C (2,700°F). If the heat cannot be redistributed, then the day side may reach values of 2,300°C (4,200°F), while the night side could sit as low as -220°C (-360°F). This would divide the planet into a lava ocean on the day side and a black rocky landscape on the night side.

The day-side boiling rock may give the planet a tenuous atmosphere. This would be a far cry from our own air, as it would be a gas of vaporised rock. As this rocky gas rose in the atmosphere, it would cool and solidify into a hailstorm of pebbles. The pebble type would depend on the temperature, with different rock compositions solidifying at different heights. The effect would be a giant fractionating column; a planet-sized version of the equipment used to separate out the different components of crude oil using their varying condensation temperatures. Modelling this process at Washington University in the US, planetary chemist Bruce Fegley, Jr described, 'Instead of a water cloud forming and then raining water droplets, you get a "rock cloud" forming and it starts raining out little pebbles of different types of rock.' CoRoT-7b appears to lack both a solid surface, and a remotely moderate weather system.

Two years after the announcement of CoRoT-7b, the Kepler Space Telescope confirmed its own first rocky planet find. The world was Kepler-10b. It had the density of Earth-like rock on an orbit of 20 hours. Another two years later and a similarly smouldering rocky Kepler-78b was found, orbiting its star in just 8.5 hours. This class of planet was a subset of the hot super Earths (and smaller) worlds that appeared to have an Earth-like composition but an orbit measured in days. Their molten, potentially volcanic terrain was nothing like our planet. These lava worlds might not have been common, but their numbers suggested hell was to be found in the Galaxy. But it is not only terrestial worlds that have alien natures stemming from their bizarre compositions.

White planet

Astronomers first observing Gliese 436b were initially struck with a similar problem to that encountered with 55 Cancri e; the measured radius did not seem to agree with the mass for any of the usual planet interiors. The planet was 33 light years away in the constellation of Leo, the Lion, and closely orbiting its star in 2.5 days. The mass was found to be 30 per cent larger than Neptune at 23 Earth masses, while its radius was roughly equal to Neptune's at just under 4 Earth radii. This made the planet too dense to have a large hydrogen and helium gas atmosphere, but not sufficiently dense to be rocky. Instead, the most likely solution seemed to be hot ice.

Like the watery solution to 55 Cancri e's composition, it seems too bizarre that ice could exist in the blazing heat of a planet 13 times closer (7 per cent of the distance) to its star than Mercury. But compression to high pressures by the planet's very large mass can create strange forms of ice that stay solid even at surface temperatures of more than 300°C (570°F).

For the planet to have acquired the ice for its bulk, Gliese 436b must have formed beyond the ice line, then migrated inwards. Its envelope of gases must then have evaporated in the star's heat to leave a thin atmosphere around the ice and rock core, or a layer of the equally strange supercritical water.

Just as the case for Gliese 436b seemed wrapped up, new observational data flung it wide open again. The data for the planet's size was analysed afresh and Gliese 436b was found to be 20 per cent larger than previously suspected. This made its interior consistent with a more regular Neptune-like gas giant, enveloped in a thick atmosphere.

Yet the planet was not done with surprises. Neptune's striking azure blue comes from the methane gas in its atmosphere; a molecule made from a carbon atom bonding to four hydrogen atoms. Yet when the Spitzer Space Telescope observed the planet, it saw carbon monoxide, but very little methane. This was mysterious since a gas giant's thick atmosphere is expected to have plenty of hydrogen to form methane with any carbon atoms. While oxygen is also present, the carbon should preferentially want to form methane at the temperatures expected in the atmosphere. Instead, the carbon had bonded with oxygen to leave the atmosphere with 7,000 times less methane than expected. Why would this happen?

Various suggestions were proposed to explain this phenomenon. Perhaps there was methane in the atmosphere, but its signature was diluted by particularly high quantities of other, heavier molecules. Yet that should lead to a denser planet than had been measured for Gliese 436b. Then, in 2015, a new solution was proposed: perhaps the hydrogen simply was not there.

The idea was that Gliese 436b had begun with a normal mix of hydrogen in its atmosphere, but the close proximity of the star had evaporated the gas. Being the lightest element, hydrogen would have been the most easily stripped of all the planet's gases. It would have exited, but the planet's gravity would have held on to the heavier elements. Without hydrogen to form a bond, the atmospheric carbon linked with oxygen to produce carbon dioxide. Meanwhile, the departure of hydrogen would have left behind an atmosphere most abundant in helium. The new helium planet would therefore have morphed into a world unlike those in our Solar System.

Due to the numeracy of Neptune-like planets close to their star, helium planets could be quite common. Rather than losing all their atmosphere or retaining it, hot gaseous planets may selectively lose hydrogen as Gliese 436b does. This gradual stripping would not be fast, taking about 10 billion years; about twice the age of our Solar System. Helium planets would then be the old wanderers in our Galaxy. They would not be blue like Neptune; instead the helium in their atmosphere would appear white.

The strangeness of the exoplanet worlds continued to expand, as was noted in 2015 by planetary scientist Sara Seager, co-author on the paper suggesting Gliese 436b's helium atmosphere. 'Any planet one can imagine probably exists, out there, somewhere, as long as it fits within the laws of physics and chemistry,' Seager said. 'Planets are so incredibly diverse in their masses, sizes and orbits that we expect this to extend to exoplanet atmospheres.'

It was a statement that would shortly prove true for 55 Cancri e.

The planet without air

Later in 2016, members of the Cambridge group that had proposed the volcanic landscape for 55 Cancri e were poring over their data from the Spitzer Space Telescope. Instead of comparing the dip in heat radiation as the planet vanished behind the star, this time they followed the planet's dim heat signature through the whole orbit. With one side perpetually facing the star, 55 Cancri e showed its night side to the telescope mid-transit across the star's surface. As it circled to duck behind the star, the planet turned to expose its day-side face. What the results revealed were two completely different temperatures.

The day side of the planet was around 2,500°C (4,500°F), between the minimum and maximum average values that the astronomers had previously seen. The night side was 1,400°C (2,600°F), lower at about 1,100°C (1,900°F). This still made the night side extremely toasty, but the huge difference indicated that this planet was very bad at sharing heat. On 55 Cancri e, the hot stayed hot and the cold stayed cold.

The huge temperature extremes suggested that 55 Cancri e had no atmosphere. Such a strong change in temperature would have driven winds in an envelope of gases, circulating the heat over to the night side in the resulting gales. Perhaps the planet once had its own air, but the incredibly hot orbit had stripped it down to its rocks.

Almost contrary to this conclusion, the astronomers spotted a further anomaly. The hottest point on the planet's surface was not directly in the middle of its day side, but shifted to the east. This pointed to at least one mechanism on the planet that was capable of moving heat. A likely culprit would be the lava. Should 55 Cancri e truly be lava world, then the molten rock could flow over to the day side and shift the hotspot. Over on the night side, the rock would solidify and no more circulation would be possible.

This picture fitted the data on 55 Cancri e well: a molten lava world with no air, dual hemispheres of eternal sunburn and engulfing darkness separated by a temperature drop of over 1,000°C (1,800°F). It was not an easy destination to pack a suitcase for,* but it was complete. At least, it would have been had not the Hubble Space Telescope found an atmosphere just a few weeks before.

* Also, your suitcase would melt.

Like the outer layers of a star, atoms in a planet's atmosphere absorb light. Missing wavelengths in starlight that grazes a planet's surface act as fingerprints for the gases surrounding the planet. Due to their small size and thin layer of gases, detecting the atmospheres of rocky worlds is a difficult task. To stand a chance, the planet must transit and the star needs to be bright and close. Previous attempts to detect the atmosphere of two other super Earths had been unsuccessful, but in February 2016 that luck changed. Hubble Space Telescope data for 55 Cancri e revealed the first atmospheric fingerprints of a super Earth.

With the telltale pattern of missing wavelengths revealing an atmosphere, how was it that 55 Cancri e had a 1,100°C (1,900°F) dichotomy in temperature? Winds should whip around the planet to redistribute the heat between the scorching day side and cooler night side. One possibility was that the planet had a lopsided atmosphere.

The idea that an atmosphere would concentrate over half the planet seems even more crazy than hot ice or a gaseous liquid water. However, it could occur if the atmosphere were a gas that condensed on the cooler night side. The planet would then be surrounded by vapour on its day side, which rained out as it cooled on the night side.

An example of such an atmosphere would be one filled with vaporised rock. If part of the molten lava on the day side evaporated into the atmosphere, it would condense back out into solid rock on the night side. The fact that this is a possibility emphasises once again the incredible heat of lava worlds. Such a system could even explain the moving hot spot on the planet's day side, which could be shifted by the day-side atmosphere rather than by lava flows.

This notion was thwarted by the Hubble observations finding an atmosphere dominated by hydrogen and helium. This in itself is a surprise, since these light gases are prone to rapid evaporation. 55 Cancri e was not expected to still be holding on to these primitive gases, which adds to the planet's ever-expanding mystery. That aside, if these gases are present, they would remain gases at even very low temperatures. There therefore should have been no problem circulating between the two hemispheres of 55 Cancri e.

The only further clue provided by Hubble was the detection of hydrogen cyanide. A favourite poison in Agatha Christie novels when in liquid form, hydrogen cyanide is formed from atoms of hydrogen, carbon and nitrogen. It is possible that this gas may be messing with how the atmosphere circulates around the planet, but currently no one is sure.

Interestingly, the presence of hydrogen cyanide hints at the composition of 55 Cancri e. The hydrogen–carbon–nitrogen combination should only be strongly present in the atmosphere of a carbon-rich planet. This puts the prospect of 55 Cancri e being a carbon world back in the picture. If it is, we can add an immensely poisonous atmosphere to the property description, just in case a diamond mantle still sounded tempting.

55 Cancri e is the planet version of not judging a book by its cover. The frankly horrifying possibilities for its surface conditions demonstrate that an Earth-like size and a Sun-like star are not sufficient to make our home planet. But this is not nearly as strange as we can go. The Sun is not the only type of star in the Galaxy. For one, it is not yet dead.

CHAPTER EIGHT

Worlds Around Dead Stars

Jocelyn Bell Burnell is possibly the person most famous for not winning the Nobel Prize. In the summer of 1967, she was tramping through 57 tennis courts' worth of countryside to set up 2,048 radio antennae. The data from the resulting radio telescope was set to be the focus of the young student's doctorate thesis at the University of Cambridge. What transpired was the discovery that would make Bell Burnell one of the most famous names in astrophysics.

Two months into examining the data from the telescope, Bell Burnell noticed a strange signal. It was a radio pulse that repeated exactly every 1.337 seconds. The accuracy was so precise that Bell Burnell and her academic supervisor, Anthony Hewish, contemplated whether it might not be the result of extraterrestrial life. The regularity with which this pulse appeared challenged even the accuracy of atomic clocks, which seemed to indicate an artificial creation by an advanced civilisation. Bell Burnell and Hewish referred to the object as LGM-1, for *Little Green Men 1*.

The little green men idea was put to rest when Bell Burnell found the same signal originating in a different part of the sky. The sources were too widely spread to be a single civilisation and it was highly unlikely that identical signatures would be used by entirely separate life forms. Bell Burnell declared that she was relieved by this conclusion, since the timing of finding aliens so close to the end of her doctoral course was particularly bad. But what was out there that could rival an atomic clock in accuracy? The answer turned out to be a dead star.

A star's gravity is perpetually trying to crush it to bits, but is thwarted by the heat created from the star's burning fuel. This energy increases the speed of the star's atoms that zip around and push against the collapse. *Burning* for a star does not involve the chemical combustion of campfires, but the fusing of light atoms into heavier ones. It is a process known as *nuclear fusion*.

Due to the smaller repelling positive charge of their nuclei, it is easier to fuse lighter atoms than heavier ones. Stars therefore begin by fusing hydrogen atoms into helium. This still requires ridiculously high temperatures so that the atoms collide fast enough to overcome their electric repulsion, with that of the Sun's core reaching 15 million degrees. Once helium forms, the atom's larger mass causes it to sink to the star's centre, leaving hydrogen to continue to fuse in the outer shells. When a star runs out of fuel to burn, gravity wins. What happens next depends on the mass of the star.

For a star like our own Sun, the weighty helium core is compressed by its stronger gravity. This raises the temperature and the star begins to swell. As the outer layers expand, they cool to emit a red hue that earns the star the name *red giant*. Eventually, the temperature in the core reaches 100 million degrees and helium begins to fuse into carbon. The heavier carbon sinks below the helium to form an even denser core. Our Sun will not have enough mass to compress the carbon core to the temperature where carbon will begin to fuse. Instead, the heat from the core will blow away the outer layers of the dying Sun and leave a dense remnant around one half of the mass of the Sun, but squished down to the size of the Earth. This is known as a *white dwarf*.

For a star of more than 8 Sun masses (or *solar masses*), the ending is rather more dramatic. The greater mass crushes the core to burn carbon and onwards to steadily heavier elements. When it reaches iron, the nuclear fusion comes to a halt. Fusing iron does not release energy, but absorbs it. This means that the star gets no boost in support from the burning. With no way to keep countering the collapse, gravity wins and the star implodes. Pretty much everything gets fused in the resulting shock wave, and the star explodes in an event referred to as a *supernova*.

If the core left behind after the supernova explosion has enough mass, gravity will become an unstoppable force. The remains of the star will collapse until not even light can escape its gravitational pull. This is a *black hole*. If the explosion leaves a core of between 1.4 and 3 solar masses, then there isn't enough mass to really finish the process. Instead of a black hole, gravity compresses the core so strongly that electrons and protons within the atoms combine to produce neutrons. The result is a hot naked ember known as a *neutron star*, the densest star in the Universe.

These stellar corpses have diameters than have shrunk from millions of kilometres to around 20km (12mi), but with masses over 40 per cent more than that of the Sun. The composition from surface to core consists of steadily more neutron-rich atomic nuclei, before even the nuclei structure breaks apart to form a soup of neutrons. A sugar-cube volume of a neutron star would weigh more than 100 million tonnes on Earth, and contain the (immensely squashed) entire population of our planet.

As the radius of the star contracts to form the city-sized neutron star, its spin is retained. The result can be compared to pulling your arms to your chest while whirling around on a rotating office chair. In both cases, you and the star will spin faster.* For a neutron star that has undergone such a massive contraction, the resulting rotation time drops to a matter of seconds.

While a neutron star mainly consists of neutral neutrons, about 10 per cent of charged protons and electrons remain to ensure that the star hangs on to its magnetic field.† The collapse compresses the field, which is magnified to become a trillion times stronger than that found around the Earth. The magnetic field rips over the surface of the rotating star, and pulls the remaining protons and electrons from the crust to be funnelled along the field lines towards the magnetic poles. As they change direction to twist around the field lines, the charged particles emit radio waves along with the more

---

* I bet you just tried that. So did I.
† As we saw in Chapter 6, charged particles can both create a magnetic field and feel a force in an existing field.

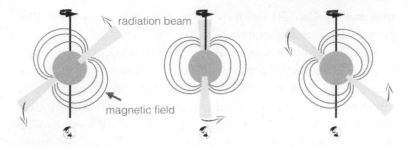

Figure 12 A pulsar is a neutron star whose radiation beams sweep past the Earth as the neutron star rotates. We observe this as a regular flash like that of a lighthouse.

energetic X-rays, Gamma rays and visible light. Where the magnetic field lines converge at the poles, the radiation narrows into beams that radiate into space, along with a wind packed with the charged particles.

The north and south poles of the magnetic field do not necessarily align with the star's axis of rotation. This is also true on Earth, where our planet's magnetic field is tilted by 11 degrees with respect to our rotation axis. This offset makes the neutron star's radiation beams swing around the star like a lighthouse beacon as it rotates. If the path of the beam passes across the Earth, our planet is flashed by a regular pulse of radiation for each rapid rotation of the neutron star. It was this pulse that Bell Burnell identified as her little green men.

When speaking to the science correspondent for the *Daily Telegraph* newspaper in 1968, Bell Burnell was asked what these strange flashing sources should be called. The science correspondent suggested 'pulsar' as a variant on the word 'quasar'; the bright but non-pulsating radio sources Bell Burnell had anticipated studying with her telescope. The name stuck and a new type of astronomical object was found.

Upon determining the source to be the rapidly rotating pulsar, Bell Burnell and Hewish renamed their first mysterious object from LGM-1 to *CP 1919*, for *Cambridge Pulsar*, with the '1919' designating its angular position eastwards in the sky. It later acquired its current official name 'PSR B1919+21', for *Pulsating Source of Radio*, with the extra '21' indicating that it was also 20 degrees north, and 'B' to specify this coordinate-naming convention.

In 1974, Hewish was awarded the Nobel Prize in Physics for the discovery of pulsars. That Bell Burnell was not acknowledged in the prize for her part in the discovery has long remained a point of controversy, although Bell Burnell herself accepted this graciously, noting that, 'I reckon I've done pretty well out of not getting the Nobel Prize!' She went on to win a large number of other prestigious awards during her career, and has been president of both the Royal Astronomical Society and the Institute of Physics in the UK. Meanwhile, pulsar discoveries became curiouser and curiouser.

Late in the 1970s, a radio source was found just a few degrees away from Bell Burnell and Hewish's discovery. Incredibly compact, it was initially suspected of being a new pulsar. However, when the sky was monitored for the signature flash, nothing was found. The source appeared to be a steady beam of radio waves rather than a pulsing lighthouse signal.

Suspecting that a pulsar's blink could be missed if it was rotating incredibly fast, further searches were conducted in March 1982. These hunted for pulsars with rotation periods down to 4 milliseconds (or 250 times every second). The fastest known pulsar at the time lay in the Crab Nebula with a rotation speed of once every 33 milliseconds. This new search would therefore pick up anything that was a substantial 10 times faster. Yet there was still no sign of a pulsar's signature flash until autumn of that year.

It was the Arecibo radio telescope in Puerto Rico that finally caught the pulsating signal. This observatory's 305m (1,000ft) dish has a movie-star history, having searched for extraterrestrial life in the adaption of Carl Sagan's novel *Contact*, and been dramatically destroyed in the climax of the James Bond film *Goldeneye*. When the huge dish imaged the sky at a blistering rate of every half a millisecond in 1982, it spotted the lighthouse flash of a record-breaking pulsar.

The new pulsar had a spin period of 1.558 milliseconds, corresponding to a staggering 642 revolutions each second. This was 20 times faster than the Crab Pulsar and created a record the pulsar would hold for the next quarter of a century.

While the discovery of a millisecond pulsar resolved the quandary of the source of the radio emission, it cracked open another can of worms. Since pulsars are continuously pumping out energy in radio waves and other radiation, over time they gradually slow down. Young pulsars therefore spin faster than old pulsars. Since the millisecond pulsar was the fastest ever discovered, this should mean that it was incredibly young. But the evidence suggested otherwise.

If the pulsar had been caught near its birth, it should be surrounded by signs of the giant supernova explosion that expelled the star's outer layers. The gas that is flung outwards from the dying star is known as a *supernovae remnant*, and is typically visible for more than 10,000 years. The Crab Nebula is the supernova remnant of the Crab Pulsar, with an estimated age of 960 years. The new millisecond pulsar should be far younger than this, but there was no sign of the gaseous remnant.

More strangely still, the pulsar appeared to be slowing down extremely slowly. Models for a pulsar's changing speed suggest that young, fast pulsars slow down quickly, and such a millisecond whip-around should be decaying rapidly over a matter of years. Measurements for this pulsar's slowdown were far lower than predicted and pointed to an age of around 230 million years. This was far, far older than any pulsar previously discovered. How could a pulsar, radiating energy into space, be both the fastest and the oldest? It would turn out that this pulsar had cannibalised its twin.

The story of millisecond pulsars begins with a pair of orbiting binary stars. While twinned by their motion in the sky, these stars are not identical and one is far more massive than the other. Size is not a healthy attribute for a star, since the extra mass causes it to burn more swiftly through its nuclear fuel supply. The bigger sibling therefore reaches the end of its normal starry life first and explodes in a supernova. Such a huge explosion at close quarters risks the smaller star being blown to bits, but if it survives it can find itself paired with a neutron star.

While the neutron star is tiny, it remains incredibly heavy. Its sibling therefore continues to feel the neutron star's gravitational pull and the two remain orbiting one another. If the neutron star's magnetic poles are aligned towards the Earth, its radio beams sweep across our planet to be detected as a pulsar. As time passes, the pulsar begins to slow. Over about 100,000 years, the pulsar's radio signal gradually weakens until it is undetectable and the pulsar falls silent. The pulsar's mass remains unchanged by this slowing, though, so its sibling star continues around its orbit. However, the sibling too is now finally reaching the end of its life.

The gravity of each star dominates a surrounding region of space called the star's *Roche lobe*. The concept of the Roche lobe is similar to that of the Hill radius for when the objects involved are similar masses. Rather than being spherical regions around the stars, the Roche lobes resemble teardrops that meet at their tapered point. At this touch point, the gravitational pulls from both stars cancel one another like the lip between two mountain valleys. Step towards one star, and its gravity would pull you towards it. Move in the opposite direction, and it would be its sibling that dragged you inwards.

As the smaller star runs out of hydrogen to burn, it swells to a red giant. The star's radius becomes so big that it overspills its Roche lobe and is drawn into the neutron star's domain. This overflow is the same mechanism that meant chthonian super Earths could form from overflowing hot Jupiters, as described in Chapter 6.

As the red giant's outer layers pour on to the neutron star, this receives a kick that causes its rotation to once again increase. As more of the red giant sibling is transferred, the neutron star's spin increases to incredible millisecond speeds. The material that hits the neutron star's surface is heated to extreme temperatures of up to 10 million degrees. Such fantastically hot material does not emit infrared, but the higher-energy X-rays. If these are detected on Earth, the paired stars earn the intermediate term *low mass X-ray binary system*.

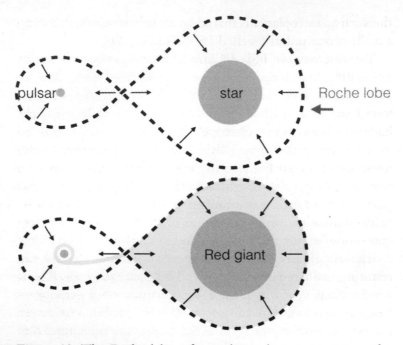

Figure 13 The Roche lobes of an orbiting binary, consisting of a regular star and a pulsar. Matter inside the Roche lobe is gravitationally pulled towards the central star. When the regular star becomes a red giant, its outer layers can overflow its Roche lobe and be drawn on to the pulsar, which spins up to reach millisecond speeds.

Eventually, the outer layers of the red giant have been sucked on to the neutron star, leaving a white dwarf orbiting a millisecond pulsar. This rejuvenation of a pulsar's spin has led to millisecond pulsars being referred to as *recycled pulsars*. Even more than the regular pulsars, the timing of millisecond pulsars is exceedingly exact – so exact, that the influence of even a tiny object is detectable.

The very first exoplanet

51 Pegasi b is frequently remembered as the first exoplanet to be discovered. Yet in truth, the hot Jupiter was only the first planet to be found around another Sun–like star. The title for

the very first exoplanet is shared between two worlds orbiting a millisecond pulsar labelled PSR B1257+12.

The discovery of PSR B1257+12 is unusual since it begins not with a brand-new telescope, but with a broken one. In 1990, the Arecibo radio telescope – the same instrument that found the first millisecond pulsar – needed repairs. Cracks had been found in the telescope's structure and no one wanted to take any risks, especially since the 90m (300ft) radio telescope at Green Bank in the US had collapsed suddenly a few years earlier due to structural failure. The Arecibo radio telescope was still usable during the repairs, but had to be held stationary in one position rather than moving continuously as it tracked an object over the night sky. This restriction greatly limited the projects Arecibo could do, resulting in a large drop in demand for the telescope's time. It was a reduction of which Alex Wolszczan, a Polish astronomer based at Arecibo, took full advantage. His plan was to survey the sky for undiscovered millisecond pulsars; a proposal that required almost one-third of the then world's largest radio telescope's time for a month. Under normal circumstances, it was a proposal that would have been dismissed. However, with demand down and Wolszczan based at the telescope, the dedication of time to this project was accepted.

The result of Wolszczan's survey was two new pulsars. The first was a pulsar in a binary, whose sibling was a second neutron star. This initially looked the more exciting system, but then Wolszczan noticed that there was trouble with the other pulsar's rotation time.

From Wolszczan's survey, PSR B1257+12 became the fifth millisecond pulsar detected. The pulsar had a rotation time of 6.2 milliseconds, making 161 rotations each second. Yet when Wolszczan tried to predict how frequently the radio beacon should be seen on Earth, he could not seem to get it right. This was particularly strange for a millisecond pulsar. Recycled by their companion, these old neutron stars suffer less from the quakes and shudders that can affect the spin of their younger and slower counterparts. One possibility was that the irregularity was due to the pulsar's orbit with its sibling. As the two stars circled one another, the distance to the Earth would vary slightly and change the arrival times of the pulses. However, no companion could be seen (in itself a strange phenomenon for a recycled pulsar), and the variations in the pulsar's timing seemed too small to be due to an orbit with a star-sized neighbour. A smaller companion also made no sense, since surely such an object would have been vaporised during the pulsar's red giant phase, or blown free of the pulsar's gravity as its mass shrank during the supernova explosion?

Wolszczan's next thought was that this was a problem with pinpointing the pulsar's position. If the pulsar's location was wrong, then the distance to the Earth would be estimated incorrectly. Such an error would change the expected arrival times of the radio pulses and thereby knock Wolszczan's predictions off base. To get a better marker, Wolszczan contacted Dale Frail at the US National Radio Astronomy Observatory's imaginatively named *Very Large Array*, or VLA for short. Based in New Mexico, the VLA consists of 27 separate radio dishes arranged in a giant 'Y' shape. These can combine their data to achieve extremely high precision measurements.

While Frail was working on pinning down the pulsar's location, the news headlines went wild: a planet had been discovered orbiting a different pulsar. The article splashed across the front cover of the journal *Nature*, on 26 July 1991, proclaimed that the 'First planet outside our Solar System' had been discovered.

The discoverers were British astronomers Andrew Lyne and Matthew Bailes, and graduate student Setnam Shemar. The planet-hosting pulsar was PSR B1829-10, a regular pulsar 30,000 light years away in the constellation of Scutum, the Shield. Variations in the pulsar's signal indicated a planetary companion about 10 times the mass of the Earth, and an orbital time of about six months.

The news gave Wolszczan mixed emotions. With planets around pulsars an option now firmly on the table, he wondered whether he had just been beaten to making history. Were planets also the explanation for PSR B1257+12's strange motion? It was a possibility that he had considered, but he did not yet have enough evidence to repeat the claim.

Frail had also read the news about the exoplanet discovery. He faxed the new updated position coordinates of their millisecond pulsar from the VLA observations to Wolszczan, joking as he did, 'Don't find any planets!' Wolszczan updated his model for the new data, and was forced to email back his reply with the news that they had just found two.

The planets were both approximately 4 Earth masses, on slightly elliptical orbits lasting 65 and 98 days, circling the pulsar on either side of Mercury's position from the Sun. By including the effects of these planets in his model, Wolszczan was able to fit the frequency of the pulsar's beam perfectly.

Before Wolszczan and Frail could publish their discovery, news of the dual planets was leaked in the popular press. On 29 October 1991, the *Independent*, a UK newspaper, hinted at the detection of two new worlds around a pulsar based on comments from Lyne. The article was cautious about the discovery, noting that 'Professor Wolszczan was not prepared to talk about his research because he feared it might prejudice the chances of it being published in a scientific journal. He also stressed that other astronomers have not yet had a chance to examine his results.' The *Independent* article was followed up by a second piece in the science magazine *New Scientist*, on 14 December 1991. Despite sounding more confident about the discovery, the short article was a surprisingly lacklustre description of the first few planets ever found beyond our Solar System. The lack of fanfare might have resulted from scepticism that such odd planets had truly been found, or been due to awaiting the scientifically reviewed journal paper. Despite Wolszczan's concerns, his and Frail's publication appeared in the 9 January 1992 edition of *Nature*, and the two planets around PSR B1257+12 were formally announced.

The journal article appeared just before the American Astronomical Society winter meeting, one of the major events in the astronomy community's calendar. It was held that year in Atlanta, in the US, and was scheduled to hear from both pulsar planet discovery groups in succession. Lyne would speak on the first exoplanet to be discovered, and Wolszczan would follow to describe their new planet pair. But Lyne's talk was not the one originally planned. Standing before his audience, Lyne admitted that there had been an error in their calculations: there was no planet orbiting PSR B1829-10. The warning alarm had been the supposed planet's six-month period. Such an exact fraction of the Earth's cycle around the Sun indicated a possible problem with the accuracy of the pulsar's known position, causing it to appear to wobble due to the Earth's own motion. Despite the care the team had taken, this error had slipped through. Once corrected, the pulsar's regular flash appeared on schedule, unaffected by the tug of a hidden companion. 'Our embarrassment is unbounded,' Lyne concluded. 'And we are truly sorry.'

Lyne had discovered the error only days before the meeting, but had attended the international conference to announce the mistake. This confession elicited shock from his audience, which turned to respect at Lyne's integrity and bravery at publicly admitting the error. The end of his talk was greeted with huge applause. This was what science was all about: trying, improving and continuously adjusting ideas to fit the new data.

For Wolszczan following Lyne's talk, this was a difficult moment. The idea that a resurrected star like a pulsar could have orbiting planets was already very hard to believe. Now the first observation of such a system had been proved false. Yet, with Frail's accurate measurements of the millisecond pulsar's position with the VLA, the two scientists had avoided the same pitfall. The planets around millisecond pulsar PSR B1257+12 were real.

The discovery stood up to scrutiny. Six months later, PSR B1257+12 was observed independently using the 43m (140ft) radio telescope at Green Bank. The oscillation in the millisecond pulsar's signal was confirmed, strengthening the claim that this planet pair was no experimental error.

Wolszczan continued to monitor the two pulsar planets over the next couple of years, searching for any extra details hidden in the arrival time of the pulsar's beacon. In 1994, he found it. There was another object orbiting the pulsar, smaller and closer than the previously discovered planets. The quick, weak signal was difficult to spot, allowing it to escape detection until then.

The announcement did spur some scepticism. Like the false signal of the first pulsar planet, this third planet's orbit matched an orbit within our own Solar System; that of the Sun's rotation. Out near the edge of our Solar System, the US *Pioneer 10* space probe had detected a fluctuation in the solar wind; a stream of charged particles that are ejected from the Sun's surface. These variations matched the Sun's rotation and the orbit of the supposed third planet. The concern was that the solar wind was dispersing the pulsar's signal as it travelled to Earth, with the strength of this weakening varying to produce a signal fluctuation that looked like a planet.

The ability of the solar wind to disperse a pulsar signal depends on the frequency of the emitted radio waves. Wolszczan therefore observed the pulsar's beacon at different radio frequencies to see if the strength of the signal changed. The whisper of the third planet remained in place: it was really there.

The new planet was just twice the size of the Moon, with an orbit taking 25.4 days. The fact that such a small planet could be detected more than 2,000 light years away was a testament to the incredible accuracy of the timing of a millisecond pulsar's beacon, allowing even tiny variations to be picked up. In principal, the technique is so sensitive that a planet the mass of a large asteroid could be detected. More than 20 years on from Wolszczn's detections, this moon-sized world remains the smallest planet ever found.

The confirmation of planets around a pulsar was great news, but it left a glaring question: how does a long-dead star have a planetary system?

Salamander planets

The most intuitive way in which a pulsar could be orbited by planets is if the system formed in the same way as our own did, at the beginning of the star's normal life. The difficulty is for the planetary system to survive the star morphing into a pulsar.

Named the *Salamander Scenario* by California Institute of Technology scientists E. Sterl Phinney and Brad Hansen, after the fire-loving mythical lizard, closely orbiting planets have to first survive being enveloped by the outer layers of their star as its size rapidly expands to a red giant. Being inside a star is not a health spa for planets. The worlds risk vaporisation and being knocked by the expanding gas to skitter deeper into the hotter reaches of the star. How far out the red giant's swollen surface extends depends on the star's mass. Around our Sun at 1au, the Earth risks being enveloped when our Sun swells in its red giant phase. The three planets around the far more massive precursor to the pulsar PSR B1257+12 would certainly be in the star's fiery belly.

An even bigger problem occurs when the star explodes as a supernova. The resultant loss of stellar material will see the star shrink to a fraction of its initial mass. The mass loss will cause a big drop in the star's gravitational pull that will most probably untie a small object such as a planet, and see the world drift away from the star. Such a fate could be avoided if the supernova blast were asymmetric, kicking the remaining neutron star towards the planets to permit a lucky recapture. Frankly, this seems exceptionally improbable, especially for grabbing hold of three planets.

A final problem is that the planets around PSR B1257+12 orbit in the same plane, which suggests that they have been fairly undisturbed since their initial birth in a disc. A shared orbital plane also makes an additional option of planet capture unlikely. A close encounter between a pulsar and another star could potentially allow planets to transfer stellar parents to orbit the pulsar after its dangerous death had occurred. Yet if the three PSR B1257+12 planets had been dragged out of their original orbits around one star to circle another, their paths should skew randomly about the pulsar. This is not what is seen. What we need is a formation option with a less violent evolution than an explosion plus recapture.

Memnonides planets

If pulsar planets orbit in a tidy, disc-like plane, then perhaps a new protoplanetary disc was created after the star had become a pulsar. This new disc could then trigger a very late generation of planet formation. In keeping with the mythological analogy, this idea became the *Memnonides Scenario*, after the Memnonides birds in Greek mythology that rose from the funeral pyre of the fallen warrior, Memnon.

A new disc fits the observation of the PSR B1257+12 system, but opens the question of the origin of the disc material. The original protoplanetary disc would have long since been dispersed, so the pulsar would need a fresh supply of planet-building dust.

One source for a new disc might be the outer layers of the red giant star thrown off during the supernova explosion. If this material fails to escape the gravitational pull of the remaining stellar corpse, it will fall back to encircle the freshly made pulsar. Assuming that the incoming material can rotate fast enough to stop falling on to the pulsar, then a disc is formed. How much of the red giant's outer layers could be recycled into a planet-forming disc is uncertain. But if the initial star were large enough, then sufficient material could fall back to create the small planets circling PSR B1257+12.

What is harder to explain is the fact that PSR B1257+12 is a millisecond pulsar, yet there is no sign of a companion star. If the outer layers from its stellar sibling were needed to spin the pulsar to sub-second speeds, the remnant of that star should remain as a white dwarf. Where is it and could this be linked with the planets?

The most promising idea for PSR B1257+12 is to blame its missing stellar sibling for the whole planet-making action. In this macabre scenario, the companion star is ripped to shreds by the pulsar and the new protoplanetary disc is formed from its broken remains.

One way to destroy the sibling is during the supernova. If the explosion is asymmetric, the newly forming pulsar could be slammed into its companion. This collision would rip apart the sibling star to form a protoplanetary disc around the pulsar. Collisions between stars are extremely rare, but pulsar planets are not yet considered common.

Alternatively, the pulsar could blowtorch its companion to pieces. Named *black widow pulsars* after the spiders that devour their mate, these cannibalistic dead stars orbit their sibling so closely that their radiation vaporises the other star to form a disc of its remains. One such stellar homicide is presently being committed by pulsar PSR J1311-3430.*

In 2012, a faint star was discovered that seemed to be varying in colour from bright blue to dull red. Its location was also a strong source of high-energy Gamma rays, but there were only intermittent intervals of radio-wave emission.

The high levels of radiation led to a suspicion that a pulsar was at the heart of this mystery. The challenge was spotting the telltale lighthouse pulse in the Gamma ray emission. Due to the high energy of this radiation, pulsars emit far fewer Gamma rays than radio waves, and this makes it hard to spot a rapid flash. Careful analysis of four years' worth of data taken by NASA's *Fermi Gamma-ray Space Telescope* eventually yielded a successful result; the colour-changing star was indeed orbiting a pulsar, the first to be identified purely by its Gamma ray flash.

PSR J1311-3430 is a 2.5 millisecond pulsar, rotating 390 times per second. The pulsar and its sibling were incredibly

---

* Pulsar names with a 'J' use a more up-to-date and precise position coordinate for their location in the sky.

close, orbiting at a separation just 40 per cent greater than the distance between the Earth and the Moon. This led to an orbital time of 93 minutes – less than the average round-trip commuter time in the UK. It was this proximity to the pulsar's lighthouse beacons that was the root of the companion star's colour change.

The side of the companion star facing its dead sibling felt the full assault of the pulsar's radiation. This pounding drove its temperature up to 12,000°C (21,600°F) – more than twice that at the Sun's surface – and turned it a bright blue. The far side of the star was a cooler red, corresponding to a temperature of just 2,700°C (4,900°F). As the star rotated about the compact pulsar, its red and blue faces were alternately visible from the Earth.

The pulsar also explained the star's faintness. The star was tiny, at only about 12 times the mass of Jupiter. Since the pulsar had a millisecond spin, the companion star must have previously donated its outer layers to recycle the pulsar's rotation. What remained was probably a helium core, possibly too light to have fully compressed into a white dwarf. Exposure to the pulsar's relentless radiation had then whittled the star down to a near planet-sized object. Its disintegrating body trailed behind the star to form a barrier encircling the pulsar. The pulsar's radio waves were scattered or absorbed by these shredded remains, allowing only the higher-energy Gamma rays to punch through and be seen on Earth. As the star continues to be vaporised away, its remains may condense into a disc around the pulsar. The result would be a lone millisecond pulsar and the starting disc for a new generation of planets.

If the pulsar is not close enough to blowtorch apart its sibling, this companion will eventual die and become a white dwarf. Tucked inside its own Roche lobe, the white dwarf can safely orbit the pulsar. Yet, this safety may not last forever. The reason is gravitational waves.

One hundred years ago, Albert Einstein predicted that the fabric of space should be rippling with waves. He pictured the Universe as a taut rubber sheet on which massive objects create curved indentations. Gravity is the result of these curves, forcing lighter objects to move towards the more deeply embedded heavier ones. As objects move, the sheet flexes to reflect their new positions, creating oscillations that travel outwards as a gravitational wave.

The first direct detection of gravitational waves was announced on 11 February 2016. It was possibly the worst-kept secret in scientific history, with news of a successful detection rumoured since the end of the previous year. The find was made by the US-based detector LIGO, which spotted the signature of two coalescing black holes. As the densest objects in the Universe, the ripples produced during a black hole merger are one of the strongest gravitational wave signals imaginable. Next in line are the vibrations from other interacting stellar remnants.

As a pulsar and white dwarf binary orbit one another, the continual flexing of space produces a steady source of gravitational waves. The energy to power these waves is pulled from the orbit motion, causing the pair to move closer together.* As the two approach, the pulsar's stronger gravitational pull will shrink the Roche lobe of the less massive white dwarf until it overspills for a second time.

While a white dwarf has not quite crushed itself into neutrons as in a pulsar, its incredible density causes it to behave differently from normal material. When it loses mass as its layers are pulled towards the pulsar, a white dwarf expands rather than shrinks. This causes more and more of the dead star to overspill on to the pulsar until it is entirely disrupted. The overflowed remains can form the disc for a generation of planets.

---

* The orbit's energy also powers tidal heating: in Chapter 5, the puffy hot Jupiter WASP-17b was tugged on to a closer circular orbit as it was heated by the varying gravitational pull from the star.

A protoplanetary disc formed from the ashes of a dead star has a couple of interesting properties. First, such discs are short lived. The shredded stellar material is pounded by the pulsar's radiation and rapidly heats, causing the disc to spread outwards. The disc density lowers until eventually it cannot form planets. Estimates for the lifetime of these recycled discs are around 100,000 years, compared with 10 million for a Sun-like protoplanetary disc. Such a short time period makes the formation of gas giants unlikely, but there is the possibility of close-in terrestrial planet formation, which could explain the trio seen around PSR B1257+12.

A disc made from the remains of a supernova explosion or the shredded body of a white dwarf is likely to have a strange composition. With nuclear fusion in the white dwarf's star predecessor largely stopping at helium, such discs would probably be carbon rich. The resulting terrestrial worlds could therefore be examples of the diamond planets mentioned in Chapter 7.

While it would not apply to the trio of worlds around PSR B1257+12, it is worth noting that there is one other very strange way to build a single diamond world around a dead star. This involves converting a star directly into a planet.

The star that became a planet

In December 2009, a pulsar was discovered with a spin time of 5.7 milliseconds, or 175 rotations per second. As a second star was needed to spin the pulsar up to millisecond speeds, the skies were searched for its sibling. This search initially yielded nothing: pulsar PSR J1719-1438 seemed to be alone.

The dead star's discovery had been made using the Parkes Radio Telescope in Australia. The 64m (210ft) dish is famous for receiving the majority of Neil Armstrong's iconic Moon-landing broadcasts. But perhaps for astronomers the telescope's greater claim to fame is as the record holder for the world's most successful pulsar finder. This pulsar was discovered 4,000 light years away in the constellation of

Serpens, the Snake. Nearly two years later, Parkes, along with the Lovell 76m (250ft) telescope at Jodrell Bank Observatory in the UK, was used to uncover its extremely strange sibling.

The slight variations in the timing of the pulsar's flash revealed that it remained in a binary with an orbit lasting two hours and ten minutes. However, the pulsar's sibling was a major lightweight, with a mass similar to that of Jupiter. So was this truly a star or a planet?

Due to the swift orbital time, the two companions were closely packed, separated by a distance of 600,000km (370,000mi); slightly less than the radius of the Sun. Since there was no X-ray emission, the pulsar's companion could not currently be overspilling its layers on to the pulsar. This meant that the companion's size had to sit within its Roche lobe, capping it at about 5 Earth radii for its current distance from the pulsar. This sibling was therefore a super Earth with the mass of Jupiter. A gas giant with a thick hydrogen atmosphere could never squeeze into a radius that small. This left the alternative of a very small white dwarf.

As the lighter versions of neutron stars, white dwarfs have a typical mass of two-thirds of the Sun packed down to the size of the Earth. To be the mass of Jupiter, PSR J1719-1438's strange companion must have donated around 99.8 per cent of its mass to its pulsar. Due to the strange configuration of matter inside a white dwarf, this would have expanded the dead star to its larger radii. Yet despite this colossal mass loss, the star had avoided being completely disrupted.

As the white dwarf's mass is transferred to the pulsar, the gravity of each changes. This alters the shape and extent of the pair's Roche lobes. If the distance between the sibling stars is just right, then this adjustment can allow the white dwarf to duck back inside its own Roche lobe and stop the overflow on to the pulsar. It is a tricky game. If the binary pair are far apart, then the white dwarf never overspills its mass on to the pulsar. If the pair are too close, then the overspill can never be stopped and completely disrupts the white dwarf.

Made primarily of incredibly dense carbon, this white dwarf world has a density above 23g/cm^3, far higher than the Earth's 5.5g/cm^3. At such values, carbon would have crystallised into a genuine solid diamond world.

This has to be one of the strangest objects in the Universe: a planet made from diamond, orbiting a sibling the size of a city, which once used to be a star.

CHAPTER NINE

The Lands of Two Suns

More than 10 years before the first exoplanet was discovered, a telescope was scrutinising the sky for a sign that our Solar System was not unique in the Universe. It was a search that struggled to be taken seriously. The scepticism was not due to scientists thinking that there were no planets orbiting other stars, but because it seemed so incredibly unlikely that they would be found with the current technology.

Planet hunting had so far focused on *astrometry*; the search for tiny shifts in the locations of stars on the sky that would indicate an orbit with a planet. The problem was that even Jupiter orbiting the Sun at 16 light years away would produce an angular movement of just 0.0000003 degrees. This was about 1,000 times smaller than the resolution of the photographic images of the sky being taken from Earth.

The closest claim to a find was the announcement of two Jupiter-sized planets orbiting Barnard's star; a red dwarf about six light years away in the constellation of Ophiuchus, the Serpent Bearer. Comparison of the star's position on photographic plates in the 1960s suggested a 1 micrometre shift in position. However, it was later shown that this movement corresponded to times when the telescope lens had been cleaned; a discovery that underlined how impossible this process seemed.

Looking for the wobble in the star's radial velocity did not seem more promising. Jupiter creates a shift in the Sun's velocity of about 13m/s over its 12-year orbit, assuming that the orbit was viewed edge-on for maximum effect. In the 1970s, a star's radial velocity could only be measured to the nearest 1km/s. This meant that the signature of even Jupiter-sized planets was impossible to detect. Even a hot Jupiter (at the time, an unimaginable entity) would be missed.

The breakthrough in radial velocity measurements was made in the late 1970s by Gordon Walker and his postdoctoral researcher,* Bruce Campbell. They proposed putting a container of a known gas between the incoming starlight and the telescope detector. Like a star's atmosphere, the gas atoms absorb specific wavelengths of light. This creates a fingerprint of dark bands overlaid on the starlight. As the light from the star shifts towards red and blue during the orbit with the planet, the gas fingerprint acts as a stationary point to measure this variation against – like the zero mark on a measuring stick. The big advantage was that both the reference point and the starlight could be recorded simultaneously, avoiding previously large errors due to the apparatus not staying completely stationary between the measurements.

Walker and Campbell originally picked hydrogen fluoride for the reference gas, since it had well-spaced absorption wavelengths that could be clearly distinguished. The disadvantage was that hydrogen fluoride was highly toxic and corrosive, and the gas container had to be refilled after each observing run. Writing an account of the new instrument in a review in 2008, Walker noted that 'Frankly, it was quite unsafe.'

Unsafe it may have been, but it was successful. The design allowed a factor of a hundred improvement in measuring the radial velocity of stars, giving an accuracy close to 10m/s. Although researchers would eventually swap the dangerous hydrogen fluoride for a gas container of iodine, the precision achieved with hydrogen fluoride would have been sufficient to detect extrasolar planets over a decade before the pulsar worlds. In fact, the failure to do so was to be a very near miss.

The hydrogen fluoride gas container had been installed on the Canada-France-Hawaii 3.6m (12ft) telescope, situated on Hawaii's Mauna Kea. Assuming detectable planets were Jupiter-analogues on orbits lasting over a decade, Walker,

---

* A postdoctoral researcher is a junior researcher who has recently completed their doctoral thesis.

Campbell and fellow astronomer Stephenson Yang surveyed 23 stars for a few nights a year over 12 years. In 1988, the teams reported the results from the first six years of observations. Seven stars hinted at a perturbation by a possible planet, one of which was a star known as γ Cephei. In what would become a standard feature of early planet discoveries, the object around γ Cephei would be dismissed due to being simply too weird.

The observations of γ Cephei revealed it to be a binary star system, about 45 light years away in the constellation of the Greek mythological king, Cepheus. The sibling stars' orbit was longer than the duration of the planet-hunting survey, so only part of their mutual loop was observed. From this section, the astronomers concluded that the stars took 30 years to circle one another, and there was an additional 25m/s wobble in their motion that repeated every 2.7 years. In their 1988 publication, this wobble was deemed a 'probable third body', yet by the survey's completion in 1995, the tone had become more cautious.

The larger of the γ Cephei pair was thought to be a giant star, expanding as it reached the twilight years of its life. These stellar elders were known for their irregular natures, susceptible to pulsations in their outer layers that look very like the wobble from a planet. The prospect of a planet orbiting a binary star system also seemed unlikely, a feeling compounded by the planet seeming to be a gas giant orbiting much closer to the star than any in our Solar System. With scepticism still running high for planet searches, this world seemed too exotic to be true. It was a wrong conclusion.

A further 10 years later, new observations of γ Cephei showed that the binary orbit actually took twice as long as previously thought, and the larger star had not yet become a cantankerous giant. This erased previous doubts and in 2003, the planet was announced as a gas giant of 3–16 Jupiter masses (depending on the unknown angle of its orbit), orbiting the larger star every 2.48 years at a distance of just over 2au. If the find had been announced in 1988, this would have been the first extrasolar planet discovered. Nevertheless, its

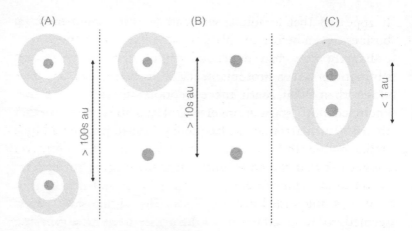

Figure 14 Formation of protoplanetary discs around binary stars. If the stars are more than about 100au apart, discs can form around both stars, although planet formation may still be disrupted if the stars are closer than about 1500au. At 10–100au, the sibling stars disrupt one another's discs. A disc may remain around the larger star, or both may be destroyed. When the stars are very close, a circumbinary disc can form around both stars.

discovery pioneered the technique that would find thousands of other worlds.

Had γ Cephei been an isolated star, the signs of an orbiting planet might have been taken more seriously. But is having a sibling a real problem for planet formation?

In the constellation of Taurus, the Bull, are several hundred newly formed stars embedded in gas clouds known as the Taurus-Auriga complex. With ages of only 1–2 million years, any of these stars planning on forming planets should be surrounded by their dusty protoplanetary discs. A survey of 23 of these young stars that orbited at least one stellar companion revealed that roughly a third were encircled by a disc. This was half the frequency seen around isolated stars.

It appeared that a sibling was bad for the planet-forming business, even before the planets had begun to form.

Exactly how bad depended on the proximity of the sibling. Very few protoplanetary discs were seen for stars closer than 30au, with the exception of a few discs that encircled both stars. Conversely, if the stars were further than 300au apart, the disc frequency seemed unaffected by a stellar sibling. In between, the discs were harder to observe, suggesting that the companion star was helping to disperse the protoplanetary disc.

If a sibling stifled protoplanetary disc formation, then it seemed to have an even stronger effect on the growing planets. Examination of stars observed with the Kepler Space Telescope suggested stellar siblings needed to be as far apart as 1,500au before planet formation was unimpeded. Planets existed where the stars were closer, but they were significantly rarer than those around stars that sat alone.

How does the companion star inflict so much damage? A protoplanetary disc forming around a young star in a binary also feels an outward pull from the stellar sibling. This distorts the disc, which bulges towards the second star. Since the disc rotates faster than the binary orbit, the bulge moves ahead of the sibling star. The sibling's gravity tries to pull the bulge back in line, creating a drag force that slows the disc. As in planet migration, the reduction in speed causes the gas and dust to move inwards and the disc contracts. The result is known as *tidal truncation* by the companion star. The same dragging effect also occurs on the Earth, due to tidal bulges raised by the Moon. The result reduces the Earth's rotation very slightly, causing the occasional *leap second* to be added to our calendar.

Truncating the protoplanetary disc cuts short the time available for planet formation. The more compact disc will sit closer to the star's evaporating heat and be accreted more swiftly. This explains why protoplanetary discs around binary stars seem to be visible for a shorter time than those around single stars. The proximity of a second star can also raise the temperature of the disc, hindering the condensation of dust to reduce the amount of planet-building material.

For the protoplanetary discs that do survive, the companion star can play havoc with the planetesimals. As the binary pair orbit one another, the second star can pull the forming planetesimals on to elliptical paths that precess like a gyroscope around the host star. The bent paths result in much faster collisions, which can cause particles to fragment rather than stick. This throttles the growth to larger planetesimals and planetary embryos.

These difficulties fuelled the suspicion that γ Cephei could not host a planet. A major concern was whether a truncated disc could even contain enough gas to form a giant-sized Jupiter planet. If it could not, then the observation was probably mistaken. In fact, estimates of the truncation point suggest that there was just enough mass to create the planet, but a second gas giant would be very unlikely.

That γ Cephei's planet is a gas giant also offers a possible solution to the rapid dispersal of the protoplanetary disc. If the disc was sufficiently heavy, then a planet could form via a disc instability and avoid the problems with destructively fast collisions. The instability might even form more easily in a binary system, as a heavier compact disc has a greater chance of fragmenting and the companion star's tug could trigger the instability.

But just when there appears to be a workable planet-formation theory for binary stars, a new planet shows up and blows it to pieces. In this case, that planet went by the catchy name of OGLE-2013-BLG-0341LBb.

The planet that bent light

Not only is the name of OGLE-2013-BLG-0341LBb more unwieldy than that of any object met so far, but the planet was also found through an entirely different method of discovery. Rather than detection from the star's wobble or a dimming in brightness, this planet was found by the gravitational bending of light.

While we do not normally think of light as being affected by gravity, Einstein claimed that rays would follow the curves in space created by massive objects.* The analogy is rolling a ping-pong ball rapidly across a sheet depressed by a bowling ball; the path of the light ball will bend around the heavy weight.

This theory was tested for the first time during the total solar eclipse in 1919. British physicist Arthur Eddington realised that the brief obscuration of sunlight by the Moon could be used to see if the Sun's gravity was bending the light from background stars. If the starlight was being bent, then the stars visible during the solar eclipse would appear in slightly different positions than at night-time when the Sun was not in the sky.

Eddington's measurements discovered a deflection of 0.00045 degrees, consistent with Einstein's prediction. The announcement propelled Einstein's theories into the international spotlight. Despite the excitement, Einstein himself was unmoved by the attention, responding to a journalist who asked how he would have felt had Eddington's observations disproved his theory with, 'I would have felt sorry for the dear Lord. The theory is correct.'

The bending of light can give away the presence of a hidden massive object, such as a dim star or a planet. It is a technique known as *gravitational microlensing*, as the hidden object acts as a lens to bend the light. In a regular lens like that in your glasses, light is bent more strongly at the edges than when passing through the middle. This allows light rays to be focused on to a single point. However, a gravitational lens acts in the opposite direction, bending light that passes close to its centre more strongly than that further away. This results in light being focused on to a ring rather than a single point, to become a bright annulus around the lens known as an *Einstein ring*. If the lens is extremely massive, such as a

---

* We met this idea in Chapter 8, talking about the generation of gravitational waves as massive objects creating ripples in Einstein's rubber-sheet universe.

whole galaxy, then the ring can be clearly resolved. In the case of a smaller lens like a star, the ring cannot be distinguished. Instead, the background star is seen to brighten and darken as it passes behind the lens, as the ring is more luminous than the background star alone.

Since anything with mass will bend light, a planet orbiting a lensing star will affect the process. This extra mini-lens creates an additional bump in the brightening and darkening of the background source. It was such a bump that revealed the presence of OGLE-2013-BLG-0341LBb.

OGLE stands for the *Optical Gravitational Lensing Experiment*, forming one of the best acronyms ever created for a galaxy survey. OGLE is organised from the University of Warsaw in Poland, and most of the observations were taken at the Las Campanas Observatory in Chile. While primarily designed to search for dark matter, OGLE has discovered nearly 20 exoplanets. OGLE-2013-BLG-0341LBb was observed while OGLE was hunting for objects in the direction of the galactic bulge, a central region of tightly packed stars in our Galaxy. This location gave the planet the 'BLG' part of its name, for 'bulge'. The '2013' denotes the year in which the observing season began, and the '0341' is just an ordinal number. The last 'L' confirms that this was an object found via lensing, to separate it out from a handful of objects OGLE spotted in transit. The capital 'B' is because the planet's star was not alone: it was one half of a very close binary.

The lensing event produced three bumps in brightness as the background star sailed behind this system: two big double peaks from each of the binary stars and a smaller bump from the planet. Analysis revealed that the sibling stars were dim dwarfs orbiting one another at 15au apart; between the distance of Saturn and Uranus from the Sun. The planet orbited one of these stars at a near-Earth distance of 0.8au and had a mass roughly twice that of the Earth. Despite the fact that the planet is a hair-width closer to its star than our own planet, it is much colder. At 10–15 per cent of the Sun's mass, the dwarf star is around 400 times less bright than the Sun. This leaves its planet a cold, dark world with temperatures likely to be about -213°C (-351°F); colder than Jupiter's icy moon, Europa. Rocky the planet may be, but it is not Earth-like.

OGLE-2013-BLG-0341LBb might not be similar to the Earth, but it proved that planets could form around a star with a very close sibling. Exactly how the planet managed to gather enough material in such a closely packed setting is open to debate, but there are a number of promising ideas. A simple solution is that the high collision speed from elliptical planetesimal orbits is less of a problem than previously thought. The gas drag may dampen the effect of the second star's pull to retain the planetesimals in circular orbits. A second option is a mechanism such as the streaming instability discussed in Chapter 2; planetesimals may gather together and collapse under their combined gravity.

Another intriguing possibility is that the binary stars were once further apart. Stars form in clusters, with many stellar neighbours packed close together. This can create star systems with three or even more sibling stars orbiting a mutual centre of mass. Such a system can be unstable, with the multitude of gravitational tugs altering the stars' speed until the group breaks and ejects a star. The result of giving the runaway star enough energy to escape causes the remaining stars to move closer together. If OGLE-2013-BLG-0341LB was once part of a triplet, then the reduction to a binary could have decreased the star separation from more than 100au to the current close orbit. The planet might therefore have formed relatively unmolested by the companion star and remained in orbit once the stars approached.

Sadly, we may never get another chance to observe OGLE-2013-BLG-0341LBb. The background star and the planetary system lens have now moved apart, and another precise alignment would be needed to see the planet again. We know this cold world is out there, but it may never be spotted from the Earth for a second time.

Despite this lamentable situation, there is another, similar system that we can repeatedly study. It is an exciting prospect, since this planet orbits a binary system that is one of the closest stars to the Earth. The only catch is that the planet may not exist.

Our nearest binary

At just over 4 light years from the Earth, the Alpha Centauri system is our closest stellar neighbour and the third-brightest star in the night sky. While it looks to the naked eye to be a single star, it is actually a triple star system consisting of a close binary and a distant dwarf star. Unlike the hypothetical trio that may have existed in OGLE-2013-BLG-0341LBb's system, the small size and large distance of Alpha Centauri's third star allows the system to be stable. The central binary stars are denoted Alpha Centauri A and B, and are approximately the mass of the Sun. They orbit one another every 80 years with an average separation of 11au; just over the distance from the Sun to Saturn. The third star is Proxima Centauri and is the closest of the three to the Earth, but sits at a faraway 15,000au from the binary.

Everyone wants there to be planets around Alpha Centauri. The system's proximity to the Earth has inspired novelists and script writers to deck the stars with civilisations from the home of the Transformers, Cybertron,* to the city that sells the best Pan Galactic Gargle Blaster in the *Hitchhiker's Guide to the Galaxy*.

The announcement in 2012 that a slight wobble had been detected in Alpha Centauri B was therefore met with great excitement. The variation in the star's radial velocity indicated an Earth-mass planet orbiting every 3.2 days. While such a short orbit pointed to a molten lava world, its existence meant that planet formation was possible in the Alpha Centauri system, and fuelled hopes that more temperate worlds might exist further out. But not everyone was convinced. The detected wobble in Alpha Centauri B was tiny – right on the edge of what was detectable. Was it perhaps a little too near that edge to be a genuine signature?

---

* In Marvel comics, Cybertron was ejected from the Alpha Centauri system to become a rogue planet. Chapter 11 considers these orphan worlds.

To extract the wobble due to a planet from the telescope data, all other influences on the star's radial velocity must first be removed. The star's own surface is a major cause of false alarms, with flares and star spots producing their own change in the starlight. In total, these effects are far larger than the variation being sought, so even a small error in their subtraction creates a false positive for a planet.

In a task that must have felt suited to a Devil's advocate, the data was analysed using a different technique for filtering out the star's noise. If the planet existed, then its signature should appear through both filter methods. But the planet vanished.

The new results did not completely kill the possibility of a planet around Alpha Centauri B, but they stamped a big question mark on this discovery. Fortunately, unlike with OGLE-2013-BLG-0341LB, the observations can be repeated to collect more data. The current difficulty is that Alpha Centauri A is moving into closer alignment with its twin from our viewpoint on Earth. This means that observers will have to wait for the stars to draw apart once again before we can ascertain if Pan Galactic Gargle Blasters are on the menu.

Searching for Tatooine

One of the most iconic views in science fiction cinema is the twin suns setting over the desert planet of *Star Wars*' Tatooine. Seen together rising and setting, Tatooine does not orbit just one sun, but circles both stars of the binary. This is known as a *circumbinary*, or *P-type*, orbit, in contrast with the previous *circumstellar*, or *S-type*, orbit, where only one star of the binary hosts the planet. While such planets were only imagined when the *Star Wars* franchise began, it turns out that they are entirely possible.

Unlike the planets circling single stars, a circumbinary planet orbits further away from the binary pair. This reduces its influence on the stars' motion, making its presence harder to detect from the radial velocity wobble. Instead, these worlds need to be caught traversing across one of the stars,

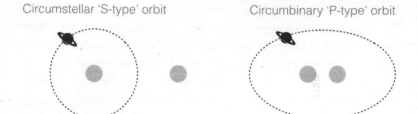

Circumstellar 'S-type' orbit   Circumbinary 'P-type' orbit

Figure 15 Planet orbits in binary star systems. Planets forming in protoplanetary discs circling one star in the binary end up on circumstellar or S-type orbits. Planets forming in discs encircling both stars orbit on circumbinary or P-type orbits.

or by their gravitational pull adjusting the period of the binary's orbit.

In 2011, the detection of Kepler-16b achieved both those feats. The system sits 200 light years away in the constellation of Cygnus, the Swan. The star pair form an incredibly close binary, with a separation of just 0.22au; less than the distance between the Sun and Mercury. Both stars are smaller than our own, at 69 per cent and 20 per cent of the Sun's mass. So close and dim are these siblings that they cannot be resolved into separate stars. Instead, their double nature is revealed by their eclipsing paths, which cause their combined light to be periodically dimmed as each star alternately dips behind the other.

In observations taken by the Kepler Space Telescope, attention was caught by three additional dips in the stars' brightness that did not correspond with the stars eclipsing. The extra dimming was a hint of a third, unseen body that was covering a little of the binary's light. These extra eclipses did not occur at even time intervals, pointing to a circumbinary orbit where the binary's rotation was shifting the transit time.

Even with thoughts of Tatooine on the horizon, it was not immediately obvious that the mystery object was a planet. It could have been a third star like Proxima Centauri, grazing past the binary as it orbited at a greater distance. This degeneracy was broken by the interloper's effect on the orbits of the binary pair. Similar to transit timing variations discussed in Chapter 6, the gravity of the third body caused slight variations in when the two stars eclipsed. The size of these changes depended on the mass of the addition. It weighed in as a planet.

This new world had the sky of Tatooine, but it was neither rocky nor hot. Close to the size of Saturn, Kepler-16b orbits the two stars every 229 days. While this gives the planet a year length close to that of Venus, the smaller stars offer scant heat, and the estimated surface temperature is -73°C (-99°F). The transit and eclipse timings revealed both the mass and radius of the planet, producing an average density of 0.964g/cm³. This is higher than Saturn (whose density is 0.687g/cm³), but far too low to make the planet terrestrial. It is possibly a hybrid world, half solid rock and ice, and half thick hydrogen and helium atmosphere.

Kepler-16b should more officially be referred to as Kepler-16 (AB) b, to reflect that it orbits both stars of the binary system. However, since there is no ambiguity when both stars are looped, the 'AB' part is typically omitted.

Are planets in circumbinary orbits common, or is Kepler-16b a rare example? At the time the planet was discovered, the latter certainly seemed possible.

As in the circumstellar case, circumbinary protoplanetary discs are subject to the dual pull from two stars. The double tug still risks increasing the speed of planetesimal collisions, grinding down potential new worlds. Additionally, if the planet formed or migrated too close to the binary pair, the changing gravitational forces as the stars changed position in their orbit around one another would vary too strongly to keep the planet's orbit stable. Instead, the planet would eventually be sent flying into the stars, or out of the system entirely.

Suspiciously, Kepler-16b sits close to this edge of stability. The stars in the Kepler-16 system are also of very different sizes, causing the larger partner to sit close to the binary's centre of mass. The most massive star therefore only moves slightly during the orbit of its sibling and planet, minimising the changes in the gravitational pull felt by Kepler-16b. This might add up to this circumbinary planet being a lucky chance.

To add to the formation problems, there is the issue of detection. Eclipsing binaries already have a periodic dimming that is far stronger than that created by a transiting planet. Star spots are a further problem, dimming the binary's light in a way that resembles a crossing world.

Circumbinary planet transits also do not occur at evenly spaced intervals, making it hard to confirm if the dip in light is from an orbiting planet. When circling a single star, the planet's transit each orbit is highly predictable, so much so that transit timing variations can be used to find hidden planets. However, a planet circling a binary must transit a moving target. The motion of both the planet and binary pair causes the iegularity of the transit clock to break. Kepler-16b's third transit across the brightest binary star was 8.8 days earlier than the time predicted based on the duration between the first two transits. This is an enormous shift, particularly in comparison to the transit timing variations caused by planetary siblings, which are typically just minutes, or at most a few hours.

To cap it all, the planet may not even transit forever. Due to the varying gravitational tug from the looping suns, the planet does not complete its orbit exactly where it started. Instead, its motion precesses, which gradually changes the planet's path. This can eventually result in it no longer transiting the stars. Models for the long-term evolution of Kepler-16b suggest that the planet will cease to transit the larger star early in 2018, and will begin to transit again in around 2042. The planet stopped transiting the smaller star in May 2014 for a predicted 35 years.

Despite this stack of difficulties, observers continued to search the skies. It was more than the intriguing nature of these double-sun planets that drove the search. The combined measurements from the transit and variations in the eclipses of the binary can give the planet properties to excellent precision. For Kepler-16b, the planet's mass and radius are each known to a remarkable 4.8 per cent and 0.34 per cent accuracy respectively. Laying hands on such precise data made the search for circumbinary worlds worth the trouble.

As it turned out, the next discoveries were around the corner. Kepler Space Telescope observations of 750 eclipsing binaries on close orbits lasting less than one Earth year were searched for a sign of a transiting planet. The search was a success: only four months after the announcement of Kepler-16b, two more circumbinary planets were discovered.

The two new planets became Kepler-34b and Kepler-35b, both gas giants with masses similar to Saturn's. Kepler-34b orbits two Sun-like stars that circle one another every 28 days, taking 289 days to complete its year around the pair. Meanwhile, Kepler-35b orbits two smaller stars, each roughly 80–90 per cent of the Sun's mass. The binary pair orbit one another in 21 days, while the planet circles around in 131 days.

All three circumbinary discoveries had orbits that aligned closely (within 2 per cent) with the orbital plane of the binary stars. This suggested that the planets formed with their stars within a protoplanetary disc that encircled both members of the closely orbiting stellar pair. If the planets had been captured, their orbits would be more randomly inclined, like those of the comets that orbit the Sun. Each of these three systems had only one planet. While this proved that circumbinary planets were possible, could a whole planetary system surround a double sun?

The answer was found the autumn of the same year. Kepler-47 is a binary with one Sun-like star and one smaller sibling a third of the size and only 1 per cent as bright. The two stars are a very closely packed pair, orbiting one another every 7.45 days at a distance of just 0.08au. Initially, the stars appeared to be alone. Their orbital times seemed unmolested by the pull of a hidden body. But the brightness of the stars revealed the shadow of another presence. In fact, there were at least two planets transiting across the larger star.

The lack of a measurable effect on the binary's motion prevented an estimate of the planets' mass. It also suggested that these could not be massive Jupiter-sized gas giants whose gravitational pull would affect the binary. Radius measurements from the transits backed this up. The closer planet was three times the size of Earth on an orbit of 50 days. Further from the stars, the second planet was larger at just over 4.5 Earth radii and took 303 days to lap the binary. They were gas planets, but not as big as Jupiter.

The proximity of the inner planet meant that it would be a hot mini Neptune, probably about half the mass of our own Neptune with a thick atmosphere. The second planet was closer in size to Neptune with a more tepid climate. While neither world offered a surface from which to watch the suns set, the presence of at least two planets was evidence that whole systems of Tatooine-like worlds might be possible.

By the start of 2015, a dozen circumbinary worlds had been found. Given how tricky these planets are to detect, it is likely that 10 times as many Tatooine-like worlds are missed in observations. Double sunsets may not be commonplace, but nor are they incredibly rare. However, there were discoveries of both circumbinary and circumstellar worlds that made even Tatooine look mundane.

Methuselah

Kepler-16b is frequently cited as the first Tatooine-like world, but it was not the first circumbinary discovery. Like the earliest exoplanet finds, the first known circumbinary planets orbited dead stars.

A year after Wolszczan and Dale's discovery of planets around a pulsar, a millisecond pulsar was observed 12,000 light years away in the constellation of Scorpius, the Scorpion. The millisecond spin pointed to the existence of a stellar sibling, and variations in the pulsar's lighthouse flash were examined for evidence of a companion. The discovery was a white dwarf star and an additional circling planet.

This planet was nothing like the super Earth-sized worlds Wolszczan and Dale had discovered. This was a massive gas giant, two-and-half times the size of Jupiter. Rather than orbiting close to the pulsar, it looped both the pulsar and the white dwarf stellar sibling at a distance of 23au – between the orbits of Uranus and Neptune in our Solar System.

The find tore holes through all the ideas of pulsar planet formation. Protoplanetary discs around pulsars were thought to come from a shredded companion star, yet this pulsar's sibling was still evidently present. Even if part of its mass could have been ripped away to form the disc, how could such a massive planet form so far away? The planet's orbit was also inclined relative to the orbit of the two dead stars, suggesting they had not formed together. All this pointed to one conclusion: this planet had been captured.

The pulsar and its white dwarf sibling did not sit in an isolated piece of the Galaxy. Instead, the pair resided in an ancient cluster of stars. When a gas cloud collapses to form a star, it does not birth just a single sun. Instead, the dense pockets of gas fragment into a close group of stars known as a stellar cluster. Clusters range in size from mere tens of stars to ones containing hundreds of thousands of new suns. Forming from the same natal cloud, this stellar family all has the same age, although differences in mass cause their life expectancy to vary. Clusters typically drift apart over time, leaving widely spaced single stars, binaries or small groups. However, the oldest and largest star clusters contain millions of members and retain their densely packed collective. These are known as *globular clusters* and were formed when our Galaxy was still young. It is in one of these ancient mammoths that PSR B1620-26 resides.

The globular cluster of PSR B1620-26 is named *Messier 4*. It is one of the easier clusters to spot with a small telescope, appearing as a fuzzy, moon-sized ball close to the bright star, Antares. The cluster has a mass of about 70,000 solar masses in a region 75 light years across, and is thought to be about 12.8 billion years old. If the new pulsar planet originally formed around a young star, then it would be approximately the same age as the cluster and one of the oldest planets ever discovered. This gave the planet the nickname *Methuselah*, after the grandfather of the biblical Ark-building Noah, who was reported to have lived for a rather unlikely 969 years.

Life within the close-knit star cluster is thought to explain the existence of a planet around a binary of dead stars. The planet could have formed in a normal protoplanetary disc around a young star. During its lifetime, the planetary system moved through the densely packed cluster core. The stars in this region become so close that interactions between stellar neighbours changed from being extremely rare, to very common. In this environment, the planet's host star drew close to a binary whose stars had already passed the end of their life to become a pulsar and white dwarf. The trio of stars interacted and the white dwarf was replaced by the planet-hosting star. The pulsar was then orbited by a regular star while the planet was jostled outwards to circle the new binary pair.

Finally, the planet's original parent star reached the end of its life and became a red giant. The giant's outer layers overflowed on to the pulsar, stripping the giant star to leave behind a new white dwarf. This process did nothing to dislodge the Jovian world, since the star was too small to explode as a supernova. Instead, it became a planet with two dead stars rising and setting in its sky. PSR B1620-26b turned out not to be the first circumbinary born from a star's death. In fact, it was not even the most strange.

The stars with one body

In the constellation of Serpens Caput, the Serpent Head, 1,670 light years away, is a binary of living and dead stars. This is NN Serpentis; a sibling star system consisting of one white dwarf paired with a regular red dwarf star. The stars orbit one another every three hours and seven minutes at a minute separation of 0.004au. It is a proximity so close that astronomers believe the red dwarf once lived inside its sibling.

While the death of a Sun-like star into a white dwarf is less violent than for its more massive counterparts, it can still be a dangerous time for a sibling. As a dying star expands to a red giant, its outer layers may not only spill on to a close companion but actually envelope it. Like a giant egg with two yolks, one star now sits within the body of the other. The pair are now said to have a *common envelope*. Inside another star, the orbit of the enveloped star is dragged inwards by the surrounding stellar material. As the orbit shrinks, energy is released that throws out the engulfing layers and exposes the white dwarf core of the red giant star.

Hidden within the folds of a red giant, stars in a common envelope are extremely difficult to observe. However, binaries of a white dwarf with a very closely orbiting companion are suspected of having once shared an envelope. Such is the case with NN Serpentis, whose two stars would have been dragged together during the white dwarf's bloated red giant phase. Being swallowed whole does not sound conducive to hosting a planetary system. But once again, the Universe did not care.

Like Kepler-16, the NN Serpentis pair form an eclipsing binary, where the stars appear to transit one another as seen from the Earth. Slight variations in when the stars eclipsed forced astronomers to conclude that – as improbable as it seemed – the stellar pair were not alone.

The small variations in the orbit of NN Serpentis were due to two planetary companions. These were massive gas giants, seven and two times the size of Jupiter. They orbited the pair of tight-knit stars much further out at 3.5au and 5.5au, taking 7.7 years and 15.5 years respectively to complete the circuit. Did these planets form with the binary stars, or were they born in the wake of the common envelope being thrown outwards?

The first option suffers from the problems of retaining a planet through a stellar death. While not as violent as a supernova, the loss of the red giant's outer layers drastically reduced the mass of the binary pair by about 75 per cent. In the case of PSR B1620-26b, such mass loss was avoided by the pulsar accreting the ejected envelope. The less massive stars of

NN Serpentis would have lost this material, making it difficult to hold on to orbiting planets. It is therefore more likely that the planets of NN Serpentis are far younger than their stars.

If the castaway envelope of the red giant formed a disc around the binary stars, then a second generation of planets could begin. If this really did occur for NN Serpentis, the two planets are very young. The white dwarf in the stellar pair is extremely hot at 57,000°C (103,000°F), implying that it has had little time to cool since its formation. This puts the white dwarf's age (and therefore, the maximum age of the planets) at no more than a million years. Such a fast formation time for two gas giants suggests that they did not form via the steady gathering of planetesimals in core accretion. Instead, the planets must have been born in a more rapid disc instability in the shell of discarded stellar layers.

NN Serpentis is an extreme system: planets born in the shredded layers of stars that have shared a body. In fact, the system is so weird that we need to end with a note of caution. Two other extremely close eclipsing binaries have been observed to have similar variations in their transit times. Like NN Serpentis, these was thought to be also due to planets. But examination of the planets' orbits revealed that they could not remain stable; the worlds would swiftly crash into the stars or shoot out of the system. This suggests that another unknown event had to be responsible for altering the binary orbits. The orbits of the planets around NN Serpentis are stable, but the comparison with similar systems implies that an alternative explanation may still exist. Closely eclipsing binaries could yet hold secrets that have nothing to do with planets.

Cousins

For planets forming around only one star of the binary, the stellar sibling can act as the evil aunt that disrupts the cradle. But is it possible for the binary stars to each host a separate system of planets?

Such cousin planets are possible, but once again depend on the separation between the stars. If the stars are far apart, then planets can form around both stars as if they were single suns. If the stellar pair are closer than about 20au, the gas surrounding the binary will preferentially be drawn around the larger star. Closer still, and the circumbinary Tatooine worlds form. Cousin planetary systems have so far proved to be rare. At the start of 2016, only three, or possibly four,[*] binary stars had been found where both stars separately hosted planets. One of these trio was the particularly intriguing case of the binary WASP-94.

WASP-94A and B sit 600 light years away in the constellation of Microscopium, the Microscope. Observations of WASP-94A showed periodic dips in the starlight, revealing the presence of a transiting planet. The star was being orbited by a hot Jupiter, with an orbital time of just under four days. To find the mass of this new world, WASP-94A was monitored for the characteristic wobble in the radial velocity. This was found, but a second wobble in the star's sibling was also detected. WASP-94B was also orbited by a second hot Jupiter that did not transit the star.

The two binary stars orbit one another at a distance of 2,700au. Such a large separation should mean that the stars had no effect on the planets forming around their sibling. The presence of the two hot Jupiters might therefore have been interesting but unremarkable, if it were not for the planets' unusual orbits.

Forming out of the natal gas cloud together, the stars' spin and the orbit of the binary and planets should all rotate in the same plane and direction. Instead, there was an angle between the orbits of the two planets that caused only one of the cousins to transit. Moreover, that transiting planet was found to have a retrograde orbit, circling WASP-94A in the opposite direction to the star's spin.

[*] The question mark here is Kepler-132, where it is not clear which star the binary's four planets orbit.

We have seen such misalignments before. Chapter 5 described how a stellar sibling could flip a planet to create a hot Jupiter with an inclined or retrograde orbit. It could be that WASP-94A and B are too far apart to interfere with their sibling's planet formation, but could still turn a planet on its head. Alternatively, these cousins might once have been planet siblings orbiting just one star. A close interaction between the two planets might have booted one of the pair over to the other star. There could also be other unseen planets in the system that sent the hot Jupiter around WASP-94A on to a tilted orbit. At the moment, the family history of these cousins remains a mystery.

Worlds of many suns

As the Alpha Centauri binary and distant Proxima Centauri demonstrate, the sibling stars are not limited to pairs. It is unlikely that more than two stars would be close enough to be encircled by a single circumbinary disc, but planets on circumstellar orbits could find their skies lit with a multitude of suns.

Like the Centauri trio, HD 131399 is a triplet star system that sits 340 light years from Earth in the constellation of Centaurus, the mythological half-human, half-horse Centaur. The group is also split into a binary and single star, but that is where the similarities with our nearest neighbours end.

The single star in HD 131399 is the most massive of the three, with 80 per cent more mass than the Sun. The binary pair consist of a Sun-sized star and a dwarf that orbit around their heavyweight sibling like a spinning dumbbell. The single star and the binary pair are separated by 300au, and sitting between them is a plus-sized Jupiter world.

The Jovian gas giant orbits the massive singleton star at a distance of 82au; roughly twice the distance from the Sun to Pluto. This equates to a staggering 550 Earth years to complete one orbit. If it were possible to live on a moon around the planet, entire generations of humans would live and die during a single one of the planet's century-long seasons. The

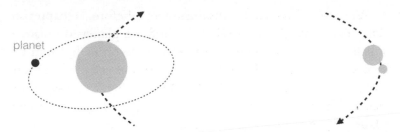

Figure 16 *HD 131399Ab orbits the largest star in a triple star system. The second two stars form a binary that also orbits with the largest star.*

planet is so far from any of the three stars in the system that its presence was not detected by transit or from changes to the stellar motions. Instead, HD 131399Ab was directly imaged.

Forming from violent impacts or rapidly collapsing gas, planets are born hot. Normally, this heat is obscured by the blaze of the star. But if the planet is big enough, young enough and far enough from its sun, then this heat signature can be detected. How hot the planet is depends on its age and mass. Assuming the age of the system is the same as that of the star, this allows the mass of a directly imaged planet to be estimated from its temperature. Direct imaging is a rapidly growing technique in exoplanet studies, but it remains hugely challenging for even the best telescopes. At four Jupiter masses, HD 131399Ab was one of the lowest-mass planets imaged at the time of its announcement in 2016.

For about half of HD 131399Ab's multi-century orbit, all three stars would sit close together in the sky. Triple sunsets and sunrises would mark each day. During the second part of the year, the stars would appear to draw apart until the rising of the primary star would coincide with the setting of the smaller binary pair. For approximately 140 years, the planet would have no real night-time. Yet despite always having a sun (or two) in the sky, HD 131399Ab would not be brightly lit. The distance to these three stars is so great, that even the massive primary would be about 1/600th the brightness of the Sun; appearing as a small but very bright point of light. The binary pair would form a double but even fainter glow.

Writing for *Slate* online magazine, astronomer and science writer Phil Plait speculated on the course of a civilisation developing on such a world. Would such a civilisation ever assume a Ptolemaic model with the heavens rotating about its planet, when an obvious counter example sat as binary suns in its own sky?

So far out, it is unlikely that HD 131399Ab formed near its present orbit. It is possible that the system has other, unseen planets that scattered the gas giant on to its current path. Alternatively, HD 131399Ab may have been a circumbinary world around the dumbbell binary, and was scattered either by the stars or a planet. The bizarreness of this situation further demonstrates that stars are excellent planet factories under any conditions.

In the constellation of Crater, the Cup, 150 light years from Earth, is a quadruple star system. HD 98800 consists of two binary star pairs orbiting one another at just 50au like, waltzers in a fairground ride. Such a close distance is the equivalent to four suns being packed into the Solar System between our Sun and the Kuiper belt. While no planets have been detected in the system, one pair of binary stars is surrounded by a circumbinary debris disc.

Debris discs consist of the dusty fragments formed during the planet-forming collisions, like sawdust in a furniture factory. Unlike protoplanetary discs, they appear at the end of the planet-formation process. The warm dust allows the discs to be spotted from their infrared radiation by instruments such as the Spitzer Space Telescope.

Around this unusual binary, Spitzer revealed a debris disc in two halves. The outer section sits nearly 6au from the encircled binary pair, while the inner part stretches between 1.5 and 2au. Between these two sections was a gap.

This space could indicate an unseen planet. The hidden world would mop up the smaller rocks in its orbit, creating a cleared section in the debris. This is not the only option, however, since the collisions that produced the debris might never have formed a planet-sized world. Instead, the largest objects would be asteroid-sized rocks. The gap might then have resulted from the combined gravitational tugs of the four close suns. However, if the planet does exist, it would see four suns in the sky. Two of these would be slightly further then the Sun from Earth. The other two would sit at roughly the orbit of Pluto.

Even more complex systems may still be out there. In the constellation of Ursa Major, the Great Bear, 250 light years away, is a five-star system. The quintuplets are split into two sets of binaries and a single star. Meanwhile, the constellation Gemini, the Twins, contains the bright star, Castor, which shines with the combined light of six stars. While no planets have been observed in these systems, they would have skies that even fiction has not imagined.

CHAPTER TEN

The Planetary Crime Scene

In 2006, the Solar System lost a planet. Ten years later, there were rumours that it had gained a new one. It is an upheaval typically blamed on US astronomer Michael Brown.

In truth, neither event involved the entrance or exit of worlds orbiting the Sun. Instead, they were the result of the growing plethora of objects found in the distant reaches of the Solar System.

The existence of the rocky band of leftover planet parts beyond Neptune has long been known. The Kuiper belt surrounds the orbit of Pluto, but it was initially assumed that its members were considerably smaller. Brown's team helped to put an end to this view with the discovery of Eris, Makemake and Haumea, all of which have sizes comparable to Pluto. Faced with the prospect of potentially having to add thousands of small planets to our Solar System list as observations improved, the International Astronomical Union (IAU) introduced a split in the population. Pluto – along with Eris, Makemake and Haumea – was reclassified as a *dwarf planet*.

The justification for this change involved more than the practicality of listing many new planets. These most recent discoveries were different from both our terrestrial and gas giant neighbours. The planets from Mercury to Neptune are local monarchs of their region of space. With the exception of the moons that are tied to the planets, there is nothing of comparable size sharing their orbits. On the other hand, Pluto's position within the Kuiper belt has it surrounded by objects that scale from Pluto-sized downwards. The source of this difference is that planets like the Earth collided with or scattered all nearby objects as they formed. Their orbits are said to have been *cleared out*. Pluto and the other dwarf planets failed to clear out their orbits, making their environments very different from those of larger worlds.

A second example within the Solar System where clearing out has not occurred is the asteroid belt. Nestled between these few million rocks is the 900km (560mi) Ceres. First discovered back in 1801, like Pluto, Ceres was considered a planet. Once the asteroid belt was revealed, Ceres was found to be surrounded by objects approaching it in size, including the asteroids Pallas, Vesta and Hygeia, which are all larger than 440km (270mi) across. In contrast, the largest known object that currently crosses Earth's orbit is the asteroid Ganymed (not to be confused with the Jovian moon Ganymede), which is about 41km (25mi) across, a factor of 300 smaller than the Earth. In fact, even Ganymed's existence is not a fair comparison, since most such Near Earth Objects were formed in the asteroid belt and only later scattered inwards. Due to the crowd it hangs out with, Ceres was classified as a dwarf planet along with Pluto and the large trans-Neptunian objects.

The outer dwarf planets' crowded orbits are not the only feature that makes them differ from their planetary cousins. Their paths around the Sun are not circles, but are elongated into ellipses and tilted out of the plane of the other planets.

As Kepler determined in the early seventeenth century, all planetary orbits are actually elliptical rather than perfect circles. Exactly how elongated the orbits are is measured by their *eccentricity*. The Earth's orbit around the Sun is so close to circular that it has an eccentricity of just 1 per cent. Mars's eccentricity is about 10 per cent, while Jupiter's is 5 per cent. These low eccentricities are not surprising. Planet formation is most efficient when planetesimals collide slowly, rather than with fast impacts that lead to bouncing and fragmentation. Slower encounters occur more frequently between rocks on close circular orbits that move at similar speeds.

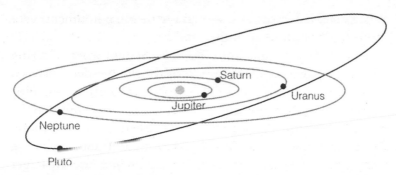

Figure 17 While the planets in the Solar System orbit in a plane on near circular orbits, Pluto's orbit is a tilted and elongated. The degree of elongation (how much the circle is squished) is measured by the eccentricity and the tilt by inclination. Pluto's orbit has an eccentricity of 25 per cent and inclination of 17°.

In contrast to these larger worlds, Pluto has an eccentricity of 25 per cent and its orbit is tilted by an angle of 17 degrees out of the plane of the Solar System. Due to this elongated path, Pluto's distance from the Sun varies so widely that its orbit crosses that of Neptune to make the dwarf planet the innermost of the two worlds for 20 years in its 248-year loop. The last time this happened was on 7 February 1979, and Pluto remained closer than Neptune to the Sun until 11 February 1999.

Due to the difficulty of building large bodies on elliptical paths, celestial objects with eccentric orbits are unlikely to have formed in their current orbits. Instead, something must have given them kick. 'It's kind of like looking at a murder scene, like those people who examine blood spatter patterns on the walls,' describes Stephen Kane. 'You know something bad has happened, but you need to figure out what it was that caused it.' In the case of Pluto, Eris, Makemake and Haumea, the culprits are Neptune and the gas giants.

The closer Kuiper belt objects such as Pluto and Haumea are locked in resonance with Neptune. Pluto and Neptune are in a 3:2 resonance, with Neptune completing three orbits of the Sun in the time it takes for Pluto to orbit twice. Due to this pattern the two worlds' orbits can cross safely without the planets colliding, as they always pass one another at the same (non-impacting) points in their circuits. Haumea and Neptune are probably in a weaker 7:12 resonance. These resonances have caused the orbits of the smaller dwarf worlds to stretch and tilt through the same Kozai–Lidov mechanism that can produce a hot Jupiter.

A little further out, Makemake is not in a resonance with Neptune. Instead, its inclined orbit is thought to be the result of being kicked outwards by the gas giant planets in the late stages of the Solar System's formation. Meanwhile, the strongly elliptical and inclined path of Eris is from a boot from Neptune that scattered it on to a highly bent path.

Beyond Eris and the edge of the Kuiper belt, the influence of Neptune ends. Far, far further out lies the Oort cloud, where a collection of small rocks teeters on the brink between the Sun's gravitational pull and the greater tug of the Galaxy. In between, there is nothing. Or so it was thought.

In the autumn of 2003, Brown was teaching at the California Institute of Technology, giving a lecture entitled 'The Edge of the Solar System'. He had held the lecture many times before and had always concluded by stating that nothing existed beyond the Kuiper belt. That year, Brown paused. 'But I'm not sure I believe this anymore,' he recalled telling the students in an article on his blog.

That morning, Brown had been reviewing images of the sky taken the previous night and noticed a slow-moving object. A few weeks later Brown, together with astronomers Chad Trujillo and David Rabinowitz, confirmed the new dwarf planet. Observed at around 100au, this new mini-world was the most distant object ever seen in the Solar System.[*] Moreover, its orbit was far from circular.

---

[*] Although the existence of long-period comets tells us that the Oort cloud is there, it is too distant to observe directly.

The dwarf planet's orbit was bent into a thin oval with an incredible 85 per cent eccentricity. This causes its distance from the Sun to vary vastly, moving between its closest distance at 76au to a far-flung 936au. Its distance from the Sun earned the dwarf planet the name *Sedna*, after the Inuit goddess who lives at the bottom of the freezing Arctic Ocean.

Sedna's eccentricity is notable, but it is its distance from the Sun that makes the dwarf planet a real anomaly. Even Sedna's closest approach is well beyond the Kuiper belt, which sits at 30–50au. This means that Sedna is out of reach of the influence of Neptune.

When an object is scattered by a larger body, it can be sent off on an elliptical path to create an eccentric orbit. Like a circle, the ellipse forms a closed loop so that the object will eventually return to the point where it was scattered. Every object on an eccentric orbit should therefore pass by the body that kicked it at some point on its path. Sedna never comes close to Neptune, or any other planet that could have imparted an eccentric-orbit forming boot. In fact, it is so far away that the Earth is closer to Neptune than Sedna. Yet, Sedna is still much too far away from the Oort cloud to be affected by the Galaxy's tidal pull. So where is the criminal that knocked Sedna on to its bent orbit?

At the end of March 2014, the mystery doubled. A second dwarf planet had been discovered on a similar highly eccentric orbit to Sedna, far away from Neptune's grasp. First observed in November 2012, this planetoid uses the moniker *2012 VP$_{113}$* until its orbit has been mapped out and the IAU can issue a more accessible name.

The elliptical orbit of 2012 VP$_{113}$ starts even further out than Sedna, stretching from 80au from the Sun to almost 450au. Like Sedna, it is on a path that does not approach any object that could have scattered it, so how did the dwarf planet get there?

At present, the definite origins of Sedna and 2012 VP$_{113}$ remain a mystery. However, there are three main possibilities. The first is that the dwarf planets were scattered not by another planet, but by a passing star.

While objects within the Oort cloud can be perturbed by stars, Sedna and 2012 VP$_{113}$ are still too close to our Sun to be affected by our nearest stellar neighbours. However, in the Solar System's infancy, our Sun was not so alone.

Stars are born in clusters that normally drift apart as the natal gas cloud disperses in the heat from the young stars. In these early years, our Sun's cluster companions would have been much closer than stars today. It is possible that the cluster exerted a sufficiently strong gravitational pull to boot Sedna and 2012 VP$_{113}$ on to their eccentric orbits. If this is true, then the two dwarf planets might belong to a much larger number of objects that form an *inner Oort cloud*. This hypothesised region is now out of reach of the influence of the giant planets and of passing stars, but is a relic of the solar system's younger years.

If the Sun's cluster stars did not scatter the dwarf worlds, then a second candidate is a planet's wild youth. Sedna and 2012 VP$_{113}$ may be out of reach of Neptune today, but the planet is not thought to have formed in its current location. After the evaporation of the protoplanetary disc, the four gas giants shuffled their positions and scattered planetesimals around the Solar System. During this period, Neptune may have been knocked on to a more eccentric orbit that swung further from the Sun. As its gravitational influence extended outwards, the gas giant could have kicked Sedna and 2012 VP$_{113}$ on to their distant paths. The outward scattering of planetesimals and dwarf planets would eventually push Neptune back on to a circular orbit, leaving it out of reach of the two dwarfs.

A push from a star and from a younger Neptune are both feasible explanations for the origins of the two eccentric dwarf worlds. But there is a third option. Our solar system may have another planet.

The allure of the idea that our Solar System may contain more planets has delivered the goods before. In 1820, French mathematician Alexis Boulevard noticed that his calculations for the location of Uranus did not match the observations. The planet was just not in the right place. The discrepancy could be explained if there were another massive object tugging on Uranus's motion. This object turned out to be the planet Neptune, which was observed for the first time at the Berlin Observatory in 1846 based on coordinates predicted from Uranus's deviant position.

A more serendipitous process led to the discovery of Pluto almost a century later. At the beginning of the 1900s, wealthy American businessman and astronomer Percival Lowell calculated that the orbits of Uranus and Neptune were also being affected by another body. While he failed to find such a missing object in his lifetime, the observatory he founded did discover a likely candidate.

Clyde Tombaugh was a Kansas country boy who constructed his own telescopes on his family's farm. One of these home-built instruments was nicknamed the *gazer-grazer* because Tombaugh had attached it to a lawnmower for mobility. In 1928, Tombaugh used his telescopes to make detailed drawings of Jupiter and Mars, and sent these to the Lowell Observatory in Arizona. The drawings resulted in a job offer from the observatory, and Tombaugh joined the hunt for Percival Lowell's 'Planet X'.

Two years later, Tombaugh spotted Pluto. This supposed ninth planet initially seemed to fit the bill for the missing force that was perturbing Uranus and Neptune. However, when Pluto's mass was determined in 1978, it was found to be too small to have the necessary effect on the orbits of its giant neighbours. The true resolution to the problem came from the NASA *Voyager 2* space probe encounter with Neptune in 1989, which found that the mass of Neptune had been underestimated by 0.5 per cent – enough to dissolve the problematic predictions of the orbits. Pluto's meagre influence on its environment saw our ninth planet being demoted to a dwarf planet around 20 years later.

But could Lowell's idea have been right all along? Was there actually a mysterious Planet X in our Solar System that was responsible for the eccentric dwarf planet orbits? Such a planet would have to be massive enough to scatter Sedna and 2012 VP$_{113}$ on to their bent orbits, but far enough away that we have yet to observe it directly.

One way to tell if we have a hidden planet is to measure the location of the Solar System's centre of mass. As described earlier, the centre of mass between two orbiting bodies is the balance point between their gravitational pulls. We compared this to the pivot position of a pencil lying across your finger that has a different-sized eraser at either end. Both bodies orbit the centre of mass, which sits closer to the more massive body.

When a system has many bodies, the centre of mass is where their combined forces cancel out. Our balancing pencil is now replaced by a plate full of heavy marbles. Unsurprisingly, the centre of mass for our Solar System is very close to the Sun. Its exact location is controlled by the positions of the planets and shifts as the planets move around. If our Solar System has a planet we do not know about, then the predicted location of the Solar System's centre of mass will be wrong. Our plate would have an extra unaccounted marble.

Detecting such a mistake can be done by using pulsars. As the Earth orbits the Solar System's centre of mass, its position relative to a particular pulsar changes. A pulsar's lighthouse beam therefore arrives at slightly different times during our planet's orbit. This change in distance is tiny compared with the pulsar's own distance, but the timing of the pulsar's flash is so precise that the variation is noticeable. To measure the correct frequency of the pulsar signal, astronomers must therefore allow for the Earth's motion. Should the estimated location of the Solar System's centre of mass be wrong, the true distance between the Earth and pulsar will be slightly off and the pulsar timings will appear irregular.

In 2005, the arrival times of pulsar signals were scrutinised for evidence of an anomalous variation that would indicate a missing planet. They came up blank. There was no evidence that an unseen planet was affecting the position of our Solar System's centre of mass.

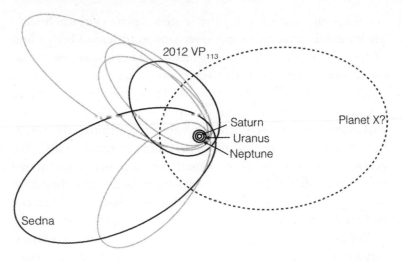

Figure 18 The orbits of the dwarf planets Sedna and 2012 VP₁₁₃
along with four other smaller objects, are tilted to suggest interference
with a mysterious Planet X.

This was compelling evidence against Planet X, but there
was still room for doubt. As in all experiments, there was a
limit to the size of gravitational tug that could be detected.
This minimum corresponded to a Jupiter-mass planet at about
200au. If Planet X existed, it had to be smaller or more distant
than such a world. Since this limit stretched far further than
any known planet or dwarf planet, it seemed unlikely that
our Solar System had a missed addition.

It was not until after the discovery of 2012 VP$_{113}$ that a
strange coincidence was noticed, which pushed Planet X back
into the limelight. The orbits of the Solar System's six most
distant objects appeared oddly aligned. Consisting of Sedna,
2012 VP$_{113}$ and four smaller bodies, this group's eccentric
orbits all overlapped at the perihelion: the point where they
passed closest to the Sun. Their elliptical paths then stretched
out in roughly the same direction.

Brown brought this problem to Konstantin Batygin, an
astronomer with an office down the hallway at the California
Institute of Technology, who worked on modelling the

motions of celestial bodies. The pair examined the observation and reached for their calculators. They concluded that such a clustering had a probability of randomly occurring of just 0.007 per cent. It was far more likely that something had aligned these orbits. The idea of Planet X was reborn.

According to Brown and Batygin's calculations, Planet X (or *Planet 9* as they referred to the prospective addition) should be approximately 10 times the mass of the Earth, or 5,000 times that of Pluto. This was no small dwarf, but a cold and distant mini Neptune. Despite its size, the planet's average distance from the Sun of 600au would mean that its influence on the Solar System's centre of mass would be missed.

The steady gravitational pull of such a Planet X would be able to slowly draw objects away from Neptune's influence in the Kuiper belt. The result would be the population of distant bodies on aligned and strongly eccentric paths.

Is Planet X observable? The answer is yes, but it is tricky.

Neptune's position was predicted by observing the deviations in the complete orbit of Uranus as it circled the Sun. By contrast, only a tiny fraction of the 1,000–10,000-year orbits of Sedna and its eccentric comrades has been seen. This makes it impossible to pinpoint the exact location of Planet X on its orbit, leaving the whole sky to be searched. It is not a small task.

The proposed existence of Planet X could resolve the bizarre locations of our outermost dwarf worlds, but how did Planet X form? With the exception of a boost in solid particles at the ice line, the density of the protoplantary disc decreases away from the central star. Twenty times further from the Sun than Neptune, it would be immensely difficult to gather the material needed to build a 10 Earth-mass planet. A passing star also remains a possibility for scattering Planet X, but the simplest option is that our gas giants did the work.

If Planet X was originally the core of a fifth gas giant, then it could have been thrown outwards during the formation of its four giant siblings. This initially seems to leave a similar problem to the one posed by Sedna and 2012 VP$_{113}$; the scattered planet should be on an elliptical orbit that returns to the scattering point. Planet X is clearly not passing close to the gas giant neighbourhood. The solution to this might be the gas drag from the protoplanetary disc. If Planet X were scattered before the gas evaporated, then its freshly elliptical path would plough across the circular-moving disc. The gas around the planet would be moving at drastically different speeds, creating a strong drag force on the eccentric world. Unlike the small Sedna or Pluto, the interaction between the gas and the strong gravity of Planet X could be enough to force the orbit into a circular route far from the Sun.

In the distant reaches of our Solar System, the gas in the protoplanetary disc would be thin. Is this enough for drag forces to circularise the path of Planet X? It is a difficulty, but there is possible precedence for such a situation.

In 2008, three gas giants were found orbiting the young star HR 8799. Situated 129 light years away in the constellation of Pegasus, the Winged Horse, the planets were still sufficiently warm from their formation to glow brightly in the infrared. It was a strong enough signature to be picked up by direct imaging, making this the first multi-planet system to be directly observed. Further observations the following year added a fourth planet to the trio, revealing a quadruple system of gas giants.

The four siblings are huge worlds. Their masses are 7–10 times the size of Jupiter, making them much more massive than anything in our Solar System. They also orbit far from their star, circling at 15–70au; a distance roughly from Uranus to well beyond the Kuiper belt. Most importantly, their orbits appear to be circular.

While the planets orbit closer to their star than do Sedna and its compatriots, their titan size makes their formation similarly tricky at their current position. There are a number of possibilities, from outward migration from the material-rich inner disc to formation via disc instability. However, it could be that interactions with another planet scattered the four siblings outwards and on to eccentric orbits. If the young planets were still embedded in the protoplanetary gas disc, the scattering planet could migrate inwards to the star and out of the way of further interactions. The elliptical paths of the scattered planets then became circular as the planets tried to plough across the remaining gas disc.

Models that replicate this circularisation suggest that it is possible, but depends on a couple of constraints. First, the scattered planet on its newly eccentric orbit has to be massive. The gas and planet have to pull strongly on one another for the drag to have significant effect. A Pluto- or Earth-sized world is just not big enough to incite a sufficient force. A dwarf planet such as Sedna could therefore not be hauled on to a circular path, but for a planet that extends into the super Earth or Neptune regime, the orbit could change. The protoplanetary disc also needs to remain thick enough around the distant worlds to provide the drag. If it evaporates before the planets are scattered, then there is no force to circularise the orbits. How likely this is remains a big unknown.

The giant planets of HR 8799 are a hint that Planet X is possible. If massive planets can truly be scattered outwards, then dragged onto circular paths, our Solar System may host a distant hidden world. It is a possibility that will keep us searching the skies around our own Sun.

Tomorrow's forecast will be 1,000°C (1,800°F) hotter than today's

With eccentric orbits being rare and difficult to explain in our own Solar System, it was reasonable to expect that such paths would be equally uncommon around other stars. This was not what was found.

Rather than a slew of regular circular orbits, exoplanets were revealed to travel around their stars in a wide variety of elongated shapes. Very close to the star within 0.1au, tidal forces from a star's strong gravitational pull ensure that a planet keeps its circular orbit. But at distances further out than the Earth's 1au position, the average exoplanet has an eccentricity of 25 per cent; more than that of any planet in our Solar System. Moreover, these eccentric worlds are not Sedna-sized dwarfs that could be easily scattered. Instead, it is the massive worlds larger the Neptune that show the most elliptical routes. Moreover, some of these examples are particularly extreme.

Snuggled into the forepaw of the constellation of Ursa Major, the Great Bear, is HD 80606. The Sun-like star sits 190 light years away and has a lone planet with a mass around four times that of Jupiter. This is a world that is truly eccentric – in all meanings of the word. HD 80606b's elongated orbit has an extreme eccentricity of 93 per cent, similar to that of Halley's Comet. Not only could this not be produced via a single scattering event, but no other planets are even seen in the system to provide the kick. The path of HD 80606b is so elongated that its closest approach to the star is just 0.03au – 3 per cent of the distance between the Earth and Sun and just four times the size of the planet. At its furthest distance, the planet approaches an Earth-like distance of 0.88au; a location still too warm for a massive gas giant to easily form. HD 80606b completes this elliptical loop in just a third of an Earth year.

The second of Kepler's laws on planetary motion states that a line drawn between the planet and star always sweeps out the same-sized area during a fixed time. We can picture this as the line being a snowplough and having a fixed quota of snow to shovel each day. For a strongly elliptical orbit, the line between planet and star shrinks when the planet passes close to the star. To meet our snow quota using the shortened snowplough, we must move much faster during our designated day. The planet similarly moves faster closer to the star, compared with when it is far away.

The result of Kepler's second law is that summer on HD 80606b lasts a day and roasts everything in sight. The planet

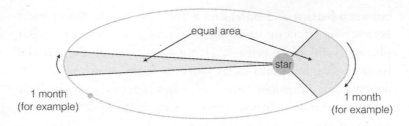

Figure 19 Kepler's second law of planetary motion: the speed of the planet varies so that a line between the planet and star sweeps out an equal area in an equal time. Planets on more elliptical orbits therefore move faster when they approach the star.

spends the majority of its 111-day year close to the location of the Earth from the Sun. Then it dives inwards to whip around the star in just 30 hours. Monitoring this fast and furious summer, the Spitzer Space Telescope recorded a temperature rise of from 500°C to 1,200°C (930°F–2,200°F) in just six hours. Even if 1,000°C (1,800°F) sunblock existed, HD 80606b summer is no beach weather. The planet's rapidly heated atmosphere expands to drive a raging storm with colossally fast winds at 18,000km/h (11,000mph). It is – to quote astronomer Greg Laughlin, who observed the planet with the Spitzer Space Telescope – 'one of the fiercest storms in the Galaxy'. Should you be able to clutch your parasol from the cloud tops of the gaseous world, you would see the star swell to 30 times the apparent size of the Sun in our own sky and increase 1,000 times in brightness.

Since HD 80606b cannot have formed on its wild eccentric orbit, we are left again with a crime scene that needs a perpetrator. Rather than neighbouring planets being too distant, HD 80606 faces the problem of not having any other planets in the system at all (or at least, none that have been spotted). With a lack of suspects, suspicions turn to its star. As it happens, this star has a sibling.

HD 80606 and HD 80607 are a wide binary system, with an average separation of 1,200au; 125 times the distance between Saturn and our Sun. While too far apart to have a very strong effect on each other's planet formation, the stellar sibling could have twisted HD 80606b's orbit though the Kozai–Lidov mechanism. In Chapter 5, this mechanism was described as a possible cause of the hot Jupiters; by alternately elongating and inclining a planet's orbit, the planet's stellar aunt can force the planet inwards.

It this occurred for HD 80606b, the gas giant could have formed at a respectable Jupiter-like distance on a regular circular orbit. Over the next 10 million years, the influence of its stellar aunt would have pulled on the planet, causing its orbit to oscillate between highly eccentric and highly inclined routes. As the planet's path morphed into an ellipse, it was forced to swing closer to its host star, whose flexing grip tidally heated the planet and drew it into a closer orbit. The result was an elliptical path, but one that only extended as far as an Earth-like distance from the Sun. The tidal pull should eventually overwhelm the effect of the star's stellar sibling to circularise the orbit of HD 80606b once again, leaving a hot Jupiter.

The interfering stellar aunt could explain some of the strongest eccentricities among the exoplanet finds. In February 2016, HD 80606b's record was broken by an exoplanet with a staggering eccentricity of 96 per cent. HD 20782b orbits its star 117 light years away in the constellation of Fornax, the Furnace. The flattened elliptical path takes the world between 0.06au and 2.5au and, like HD 80606b, this planet's star has a sibling that may have drawn out the planet's orbit. That said, there are other crime scenes weird enough to require a different explanation.

Planetary bumper cars

In the late 1990s, many Sun-like stars had been found to host a planet, but none had been discovered to be orbited by multiple worlds. The star that changed that was υ Andromedae A.

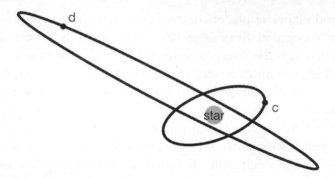

Figure 20 Planets c and d around υ Andromedae A have orbits strongly inclined with respect to one another.

Like our Sun, υ Andromedae A hosts four gas giants.* Unlike our Sun, these all sit roughly within Jupiter's orbit. These are giant giants with minimum masses of 0.7, 2, 4 and 10 times Jupiter's mass. The smallest and closest of the quadruplets is still at least twice as massive as Saturn and orbits the star in just four to five days at a distance of 0.06au.

A planetary system with four such huge planets all within the orbit of the Solar System's first gas giant was an incredible find. The fact that this was the first multi-planet system around a normal (non-dead) star made history. Yet what made υ Andromedae A truly fascinating was that its planets' orbits turned out to be completely out of whack.

The second and third planets from the star (υ Andromedae A c and d) travel on elliptical orbits with strong eccentricities of 26 per cent and 30 per cent. Compared with HD 80606b and HD 20782b, these numbers may now seem unadventurous until we consider the inclination of the orbits. Not only do the planets not orbit in the same plane as their star's spin, but they are not even in the same plane as one another. The two planets have a mutual inclination between their orbits of a steep 30 degrees. As a comparison, the angle between the

* … or three. The discovery of the outermost of these gas giants is controversial.

orbital planes of planets in our Solar System is typically less than 2 degrees. Even Mercury's orbit with its 21 per cent eccentricity has only a modest tilt of 7 degrees. The 30-degree difference sees the massive planets circling their star in wildly different directions, more like humungous comets than planets.

What could cause the strong orbital eccentricity and wildly different inclinations? The first suspicion was once again the Kozai-Lidov mechanism. As its name suggests, υ Andromedae A is indeed a binary star. Its sibling, υ Andromedae B, is a much smaller red dwarf. The pair orbit reasonably far apart, but the exact distance is unclear because the fainter sibling's motion on the sky is difficult to measure due to the brightness of υ Andromedae A. The stars could be as close as 700au or as far apart as 30,000au. However, nowhere in this range can the small υ Andromedae B have a strong enough influence to cause the complete orbital mess of the planets. The blame here was not another star.

This turned suspicions on to the planets themselves. While certainly smaller than a star, colossal worlds can provide a lot of gravitational punch. Could this be a similar situation to our dwarf planets and the distant gas giants of HR 8799, but instead of there being an outward scattering there was complete orbital mayhem?

In order to accumulate their huge masses, the planets must have begun life in the protoplanetary disc with near-circular orbits. As gas giants, they must also have been formed much further out than their current locations to include ice to bulk up their masses. The starting picture for this system would have looked similar to our own Solar System, with massive distant planets with low inclinations and eccentricities.

As their masses increased, the worlds would have begun to migrate towards the star. Close to their current positions, the gas disc then finally dispersed and ended migration. With the removal of the gas, the drag forces that encouraged the planets to keep their circular orbits also vanished. This left the planets susceptible to the close gravitational pulls of their giant siblings. While we cannot know for sure what happened next, it is very possible that planet d was given a bump.

A possible bumper car is an additional planet that was originally in the system. The gravitational pulls of planet d and the extra world accelerated the planets towards one another. Planet d was scattered on to a strongly eccentric orbit and its perturber was slingshotted into space.

Now no longer orbiting in a well-behaved circle, planet d pulled on planet c, gradually forcing its sibling to change orbit. Interestingly, the orbit of planet c is not fixed. Rather, its continual interaction with planet d should mean that it returns to a circular path, then back to an eccentric, inclined route within around 10,000 years.

These gravitational pinball games between the planets spell out an important lesson in planet building: even after formation, planetary systems may not stay the same forever. Like a virtual reality TV show, some planets may be voted off the star.

With a quarter of exoplanets on eccentric orbits, it is safe to assume that gravitational pinball is a favourite pastime around stars. Once on elliptical paths, planets experience extreme seasons such as the temperature changes on HD 80606b. Does this alone spell doom for finding a clement Earth-like world? Due to our near-circular orbit, the Earth receives a steady source of heat from the Sun. If our planet had been scattered on to a stronger ellipse, the world could have been one of scorching Mercury-like summers and freezing Mars winters. Such a world would be a far more challenging place on which to develop life.

Despite the prolific number of eccentric orbiting exoplanets, all is not lost for terrestrial worlds. Smaller in mass and with weaker gravitational pulls, these are far less likely to incite a planet pinball game. In surveys of exoplanets of around 2.5 Earth radii, orbits appear mainly circular with low eccentricities. Although we currently have fewer examples of these smaller worlds, it suggests that our pleasant circular orbit is not necessarily rare.

What happened to the fifth planet in υ Andromedae A? Ejected from its star, this world could not orbit at all. Instead, it went rogue.

CHAPTER ELEVEN

Going Rogue

A problem with mapping out the history of our Solar System is that it is very hard not to lose a planet. When the last wisps of our protoplanetary disc evaporated in the heat from the young Sun, the planets were released from the gas drag. As we have seen, their motion did not stop here. Pounded by a sea of leftover planetesimals, the orbits of the gas giants begin to move and cross. The resulting gravitational mayhem shuffled Uranus and Neptune outwards and hurled rocks around the Solar System.

While the Moon's pockmarked surface is evidence of this wild planetary adolescence, the exact details are surprisingly difficult to pin down. The issue is that Jupiter is a huge bully. Attempts to model the movement of the planets during this epoch find that our mammoth planet's huge gravity will frequently slingshot one of its neighbours into outer space. This leaves a planetary system with one less gas giant. But what if this really happened?

Modelling virtual systems that begin with five gas giants turns out to be more successful at reproducing our own Solar System than with just the big four. The extra planet forms just past Saturn with a similar size to the two icy worlds, Uranus and Neptune. During the ensuing chaos of planet rearrangement, this extra world passes slightly too close to Jupiter, which aggressively boots it from the Solar System. No longer bound to our Sun, we can never know for sure if we once had this fifth gas giant. The lost world will have sailed away into deep space. That is, the planet will have gone rogue.

The two most successful methods for hunting down planets are the radial velocity technique, which searches for the telltale wobble in the star's position, and the transit technique, which spots a dip in the starlight. The drawback is that both of these depend on the planet having a star. A rogue planet is an orphan world that does not orbit any star. There is therefore no regular, periodic effect to act as a beacon for these stray waifs of our Galaxy. This leaves two options for detection: gravitational microlensing and direct imaging.

For reasons that at first appear baffling, astronomers like building telescopes on the tops of volcanoes. This is actually because the dry and still air at the mountain tops of Hawaii provides the best view of the northern hemisphere sky anywhere in the world. The fact that these peaks are volcanic is a small disadvantage compared with the unrivalled vistas they allow of the Universe.

The Haleakala ('House of the Sun') volcano takes up the majority of the Hawaiian island of Maui. It is on this summit that the Pan-STARRS 1.8m (6ft) telescope sits. Standing for *PANoramic Survey Telescope And Rapid Response System*, the instrument aims to image the entire visible sky several times a month. This wide area results in a lower resolution than for more focused observations, but it is perfect for identifying moving objects. Quick changes in our sky could reveal an asteroid or comet that might pose a danger to Earth. The huge database of images that Pan-STARRS collects is equivalent to 60,000 smartphone photographs each night. It was in this mass of information that an oddball was seen.

While primarily designed to spot asteroids, the wealth of data Pan-STARRS produces is a valuable resource for many projects. One of these was a direct image search for very low mass stars known as brown dwarfs. These dim objects are not massive enough to burn hydrogen in their cores, but do emit a faint red heat signature. Then an object was imaged that was redder than any other brown dwarf in the data.

This redder than red source was 80 light years from Earth and designated PSO J318.5-22. The 'PSO' stands for 'Pan-STARRS1 Object', while the subsequent digits provide its sky coordinates. Comparing the dim red beacon of PSO J318.5-22's light to known stars and planets, it seemed far more similar to a young planet than other known brown dwarfs. If that was the case, where had this world come from?

With no star close enough to claim the planet, PSO J318.5-22 appeared to be free floating alone in space. However, near the rogue world was a collection of young stars known as β *Pictoris*. β Pictoris sits in the constellation of Pictor, the Painter. The group is close to PSO J318.5-22, moving at a similar velocity, and is also young. Moreover, at least two of its stellar members are known to host gas giant planets.

Comparison with the star ages in β Pictoris makes PSO J318.5-22 around 12 million years old. With the protoplanetary gas disc liable to evaporate after 10 million years, the world is a newly formed adolescent in planet terms. It weighs in at close to 6 Jupiters, putting it well below the mass of even a brown dwarf star.

A likely history for PSO J318.5-22 is that the planet formed around a star within the β Pictoris group. Ejection could happen near the end of a star's life. After the star swells to a red giant, a planet may be expelled from the system as the outer layers of the star are blown away. Alternatively, the remaining stellar remnant may lose too much mass for its gravity to keep the planet in orbit. However, both PSO J318.5-22 and the β Pictoris stars are young. It is therefore more likely that ejection happened as a result of this planet being scattered by a second planet or neighbouring star. Kicked from the group, the planet went rogue. Plausible though this sounds, it is hard to prove this theory. Do we know for sure that rogue planets have always been scattered?

The planet HD 106906b is not rogue. It orbits a star about 300 light years from Earth in the constellation of Crux, the Cross. What is difficult to explain is the incredibly large distance between planet and star.

HD 106906b is a young, super-sized gas giant with 11 times the mass of Jupiter. Like PSO J318.5-22, it was spotted via its heat glow using direct imaging. Being so small and dim, direct imaging of planets is often impossible due to the overpowering light from the star obliterating the planet's glow. For rogue or very distant planets, the absence of starlight makes this process considerably easier. Both HD 106906b and PSO J318.5-22 are also massive and young, and their bodies still smoulder hot from their formation. HD 106906b is close to PSO J318.5-22 in age; a teenager at 13 million years old.

HD 106906b's distance from its host star makes Neptune look like a hot Jupiter. It is a whopping 650au from the star, comparable to Sedna in our Solar System. Our furthest large planet, Neptune, sits at just 30au. At such distances, the protoplanetary disc would have been too thin to form a gas giant by either core accretion or disc instability. So how did this planet get to its current location?

A logical answer to this question could be provided if HD 106906b is 'almost rogue'. The planet may have been scattered by a larger planet to a very distant orbit, but just managed to stay orbiting its star. But in this particular case, the situation is complicated by two other pieces of evidence.

First, HD 106906b is the sole planet spotted orbiting the star. There is no sign of another planet that would have scattered HD 106906b on to a far-out orbit. Such a planet kicker would need to be comparable in mass to HD 106906b's significant 11 Jupiter masses, so should be detectable. There is also no binary sibling to the star that could have disrupted the planet's orbit.

Second, the star is surrounded by a sizeable debris disc. These rocky remains of planet formation were what drew observers to examine the system, not expecting to find a planet so far out. In 2014, the disc was observed to extend from roughly 20–120au; extending to just beyond the prime planet-forming real estate. A scattered planet would therefore have to be kicked right through the debris, disrupting the disc.

This led to a new idea being put on the table: could the planet actually be a stellar sibling, forming as a star by collapsing directly from the gas? If HD 106906b and its host star were a binary, there is a spectacularly large difference in mass. HD 106906b is only 1 per cent of the mass of HD 106906, whereas binary systems typically have ratios greater than 10 per cent. Was this really plausible?

A year after HD 106906b's discovery, fresh evidence changed the picture once again. New observations of the debris disc revealed that it extended much further than previously thought, running from 50au to 500au. More importantly, it was not the initially supposed undisturbed field of rocks. Instead, the outer part of the disc appeared highly asymmetric, with a needle of rubble extending outwards. Could this be the evidence that HD 106906b had indeed been scattered and ploughed through the debris?

The answer remains ambiguous. Due to the tiny size (relative to a star) of HD 106906b, a scattering event is the most physically simple. Yet, there still remains no evidence of the planet that did the scattering. Either a massive world is managing to hide from the observations, or a chance event saw a passing star give the planet a kick. The disruption in the debris disc also appears to be mainly confined to its outer parts. This could imply that a planet has not ploughed through the whole disc. Instead, the pull from HD 106906b forming at its current position could have dragged on the outer parts of the debris, leaving the more distant inner region less affected. A similar possibility is that HD 106906b was actually scattered from around another star. Ejected from its original system, the planet might have become a rogue world, but then passed close enough to HD 106906 to be snaffled and drawn into a distant orbit.

The degeneracy of the scattering and formation models could be resolved if it were definitely impossible to form a planet-sized object by a star-like direct gas collapse. Surprisingly, it turns out that this can happen.

The difficulty with forming planet-sized objects via gas collapse is that of mass. More mass equates to stronger gravity. Therefore, to produce a strong enough gravitational force to collapse inwards, the object must contain enough mass to overwhelm the outward pushing pressure of the gas. For an object as small as a planet, this means that the gas must be incredibly dense. Such densities could occur in a disc instability within a protoplanetary disc, but were not thought to ever happen in the more diffuse gas clouds that birth stars. However, that assumption proved not to be entirely right.

When a cluster of stars forms, the energy pouring from the new stars blows a hot bubble in the surrounding gas cloud. As the bubble expands, it pushes gas outwards to pile up at the bubble's rim to form a dense shell of material. These regions create the contrasting images of the nebulae, with the dark features marking the cold gas that has been compressed into shells. It was close to these shells that broken-off teardrop fragments of dense gas were seen.

The Rosette Nebula is a star nursery about 4,600 light years away. It is home to many of these tiny fragments, which gained the name *globulettes*. Formed in the outward shove of the expanding bubble, the globulettes are very high density but have masses of less than 13 Jupiters. If their cores collapsed, a planet-sized object would be born that was not attached to any star. It was a second possible method for a planet to be rogue.

There is no surefire way to differentiate between a free-floating planet ejected from around a star, versus one born as an orphan from a globulette. The best we can do is scrutinise the neighbourhood of rogues for stars that may have given up their parenting duties. So between these two methods for forming planets without stars, just how common are rogue worlds?

🪐

Although directly imaging a planet free from the overwhelming starlight is easier than for regularly orbiting worlds, the dim heat signature is still difficult to find. To attempt a survey of rogue worlds, we need to turn to the second method of detection: gravitational microlensing.

As we saw in Chapter 9 when hunting for planets around dim binary stars, gravitational microlensing does not require the light of a host star. Instead, it detects the planet as its mass bends light from a passing background star, briefly brightening its luminosity like a lens. Because it involves the chance alignment between the planet and background star, planet microlensing events are only visible for a few days and requires lucky timing to catch on. However, this is exactly what microlensing surveys are designed to do.

Two large surveys working to catch microlensing events are OGLE and MOA. It was the OGLE survey that discovered the planet orbiting within the binary star system. The second survey, MOA, stands for *Microlensing Observations in Astrophysics*, and is a collaborative project between New Zealand and Japan to hunt for dimly lit objects from dark matter to exoplanets in the southern hemisphere. Like OGLE, MOA searches for microlensing events near the galactic bulge, where the high density of stars increases the probability of a chance alignment.

The fleeting nature of microlensing events makes it essential to take observations as soon as a lensing alignment is detected. To catch these opportunities, OGLE and MOA have alert systems that spot the sudden brightening of stars and allow swift follow-up observations. In 2011, the two surveys jointly presented 10 rogue planet finds, all with a similar size to Jupiter. By estimating what fraction of the dark planets the surveys probably found, the teams put an approximate number on the rogue worlds that wander our Galaxy. Their conclusion was that there could be a staggering 400 billion rogue planets, a figure almost twice that of the number of stars.

Regardless of how they formed, the Jupiter-sized rogue worlds are not planets that we could picture standing upon. Since ejection of planets is easier if the worlds are smaller, it is nearly definite that the Galaxy is also littered with terrestrial planets without suns. These smaller bodies are currently too low mass to be detected with the present techniques. Yet even if a rogue planet is a gas giant, there is the possibility that it might harbour rocky moons.

The huge mass of the gas giant planets makes them big attractors for smaller worlds to orbit. Jupiter has at least 67 moons, with the four largest – Io, Europa, Ganymede and Callisto – ranging from two-thirds the mass of our Moon to double its mass. This makes them sizeable worlds in their own right.[*] If the rogue gas giants were ejected from their star system, it is likely that their moons would followed.

Moons may not be restricted to planets that had already acquired their satellites before ejection. The Chamaeleon complex is a star-forming region consisting of three clouds. These are descriptively named Chamaeleon I, II and III. As their names imply, the clouds reside in the constellation Chamaeleon, the Lizard, in the southern sky. Chamaeleon I has a few hundred stars, and among these an unusual disc-shaped region was observed.

Cha 110913-773444 takes its name from its coordinates within the Chamaeleon clouds. It is a free-floating object with a mass of 8 Jupiters, making it a planet-sized rogue world. Surrounding the rogue planet is a flat, dusty disc, just like the protoplanetary discs that surround young stars. If the dust in this disc later makes orbiting worlds of its own, then the rogue planet acquires one or more moons.

Although a prospective moon provides a rocky surface, would these worlds be nothing more than heat-less deserts? Without the warmth of a star, is the moon doomed to be a perpetually dark, pockmarked rock?

This is where the moons around our own gas giants offer a sliver of hope. The outer neighbourhood of our Solar

---

[*] Whether a moon could form an Earth-like environment for life is discussed in Chapter 16.

System is too cold to support liquid water on the surface of an Earth-like world. Yet there is evidence that several moons may harbour oceans under their icy surfaces. The most likely candidates for these secret seas are Jupiter's moons, Europa and Ganymede, and Saturn's moon, Enceladus. So far from the Sun, these icy orbs do not receive enough radiation to keep their oceans liquid. Instead, it is the presence of their parent gas giant that provides the heat.

Because these moons are not the lone satellite children of their gas giants, their orbits around the planet are not perfectly circular. Europa and Ganymede tug on one another and on Jupiter's innermost large moon, Io, while Enceladus is pulled by its sister moon, Dione. These different pulls result in the moons maintaining slightly elliptical orbits that cause them to feel a changing gravitational pull from the planet as they circle. The varying force flexes the moon like a rubber ball, creating heat from the friction of its interior continuously having to distort. This is the tidal heating mentioned in Chapter 7 as a way of driving volcanic activity on 55 Cancri e. It is a heat source so effective that Europa is considered to be the best bet for life in our own Solar System, outside the Earth.

While the presence of a sea certainly does not guarantee seafood, life is found on Earth wherever there is water. A subsurface ocean on a moon of a rogue planet therefore might be habitable to someone or something. But what if there is no gas giant? Could an Earth-sized rogue be a habitable shelter in the freezing bleakness of deep space?

There is no doubt that if the Sun went dark, we would be in trouble.* While the Earth gets a small amount of heat from radioactive materials and the residue warmth from its collisional formation, this is thousands of times smaller than what we receive from the Sun. Alone, this energy would not

---

* Although not for about eight minutes, since that's how long it takes the Sun's light to reach us.

be sufficient to stop our seas and oceans from freezing solid. To stand the slightest chance of life, a rogue planet needs to stay warm.

However, this comparison with the fate of the Earth if we lost the Sun may be unfair. A rocky planet sent rogue from a protoplanetary disc would not look like the Earth today. If it were ejected during the late stages of planet formation, rogue Earth would still have an early atmosphere made from protoplanetary disc material. Rather than our current mix of nitrogen, carbon dioxide and oxygen, this would be a blanket predominantly of hydrogen.

In the normal course of terrestrial planet formation, the primitive atmosphere of light hydrogen atoms is stripped by the ultraviolet rays from the young sun. Yet if the planet is ejected before its hydrogen has been lost, the coldness of outer space would make it easier to retain. As the hydrogen air cools and becomes denser, it becomes extremely bad at radiating heat and forms an effective blanket over the planet. Even at the remarkably low temperatures of outer space, the hydrogen would remain a gas and not condense on the planet's surface. The small amount of energy from the planet's radioactive rocks would therefore be trapped by the atmosphere, giving a surface temperature potentially warm enough to support liquid water.

The only catch is that the planet would need to have accrued a thick layer of hydrogen while within the proto-planetary disc, at least 10–100 times thicker than our current air. This amount is not impossible for the Earth, which could have grabbed enough hydrogen to produce an atmosphere 1,000 times greater than at present.

Of course, if rogue Earth were booted from the solar system sometime after the gas disc had evaporated, it might have already lost its primitive atmosphere. This eventuality could even still occur if our Solar System were disrupted by a passing star. The probability of such an event before the Sun moves towards its red giant phase in 3.5 billion years is about 0.002 per cent. These are not the odds to keep you up at night, but they are significantly higher than winning the National

Lottery.* The ejection of our planet might save it from being roasted in the Sun's brightening luminosity, but would anything on Earth survive a trip to deep space? With our current atmosphere, the liquid water on the rogue Earth's surface would freeze solid. This would doom human life, but what about life in a subterranean ocean like the oceans of the gas giant moons?

Without a gas giant to flex the rock and provide tidal heating, rogue Earth would have to keep an underground sea liquid with only its radiogenic heat. The prospect does not look good. Completely removed from the Sun's energy, our planet's inner sources would lead to a frosty surface temperature of just -235°C (-391°F). This is cold enough to produce a layer of ice around 15km (9mi) thick. Beneath this frozen lid, liquid water could potentially exist. Unfortunately, the Earth's total water supply is only sufficient for a global ice layer of around 4km (2.5mi) deep. All of our planet's water would therefore end up solid, with none remaining to pool as a hidden sea.

The best we could hope for is for rogue Earth's geological activity to produce hot pockets of liquid water around erupting vents on the frozen ocean floor. Life could develop in these mini pools, but it would be isolated from the rest of the planet and be exterminated if the hot pockets disappeared.

Rogue Earth is not a prospect that fills anyone with hope for a holiday destination. But does this mean that all terrestrial rogue planets are dead if they lack a primitive atmosphere? It turns out that there are four ways to improve rogue Earth's prospects: increase its water content, increase its mass, adjust its atmosphere or take the Moon along for the ride.

All these options are plausible. Terrestrial planets around other stars may have accumulated more water by forming near the ice line, or undergoing a heavier bombardment of

* So if you buy a lottery ticket, you should probably buy a spacesuit.

icy meteorites. One of the prospective compositions for 55 Cancri e in Chapter 7 was indeed a world drowning in oceans. If there is enough water to create a 15km (9mi) ice layer and have liquid to spare, the rogue planet will have a subterranean sea.

Alternatively, a more massive rogue planet will have commandeered a larger supply of radioactive elements and residue formation heat. The extra internal warmth would thin the icy lid required to support liquid water. A 3.5 Earth-mass planet with a similar water fraction to ours could have an ice layer a few kilometres thick and enough water for a hidden sea below.

Even assuming Earth mass and water content, rogue Earth might stay warmer if its atmosphere adjusts. Volcanic activity on the Earth ejects carbon dioxide into the air, which is then removed in a chemical reaction with silicate rock. When the temperature cools, chemical reactions slow and the atmospheric carbon dioxide increases. Normally, this causes the planet to warm as a result of the raised levels of carbon dioxide efficiently trapping heat in a greenhouse effect.[*] On rogue Earth, the carbon dioxide would freeze on the planet's surface, providing an extra layer of insulation that would reduce the required thickness of the ice.

A final interesting option might occur if the Moon went rogue with the Earth. Ejection from the Solar System would give the Earth and Moon a substantial shake-up. If they remained bound together, the Moon would probably find itself on an eccentric orbit. As the Moon approached rogue Earth on its new elliptical path, both bodies would feel a varying gravitational tug. In the case of a moon around a gas giant (or a planet around a star), the changing pull of the small satellite makes little difference to the huge planet. However, the Earth is a rocky world and much closer in size to the Moon than Jupiter is to its own satellites. Both rogue

[*] This is Earth's carbon-silicate thermostat, which is discussed much more in the next chapter.

Earth and the Moon would therefore flex in the changing gravitational pulls and become tidally heated.

Without additional moons to provide a countering tug, the Moon would eventually be pulled back on to a circular orbit. During this time, tidal heating of the rogue Earth could boost its energy up to 100 times above the Earth's current internal energy supply. This would last (in dwindling amounts) for around 150 million years. Generally speaking, this makes moons good news for rogue planets.

A hidden liquid ocean on a rogue planet would endure as long as the planet's internal heat would last. Over time, the planet's heat would leak away to leave a cold interior. The timescale for this turns out to be comparable to the Sun's lifetime before becoming a red giant and destroying our habitat. While this is a slightly morbid comparison, the similar time limits for Earth and rogue Earth suggest that life in hidden oceans on a rogue world has a chance of developing.

Life on a rogue world would evolve very differently from the Sun-loving surface creatures of the Earth. However, it would be wrong to claim that it would be entirely unknown. Deep in the Earth's oceans, fissures in the sea bed give rise to hydrothermal vents where sea water touches magma. These boiling-hot springs are teeming with life, despite being too deep to receive any sunlight. This may even have been where the first life on Earth began. If so, life developing on a rogue world might begin in a way that is very similar to that on Earth.

Creatures living around the deep-sea vents are known as chemoautotrophs, and they utilise the strong changes in temperature around the vents to power their biology. This process is not as efficient as photosynthesis, but a planet with no Sun might evolve organisms to better take advantage of this technique.

The theory that a rogue planet could truly support life opens the door to a couple of intriguing speculations. First, a rogue world could potentially be our nearest source of extraterrestrial life, should such a freely roaming planet pass close to our Solar System. Second, an inhabited rogue world ejected from its star system could form a delivery service for spreading organisms between planetary systems. This could be a way of life expanding through the Galaxy without requiring a separate genesis around each star system, or an advanced civilisation capable of interstellar travel.

The idea that life on Earth may have begun via microorganisms carried through outer space is known as *panspermia*. While not the most widely believed scenario for our planet, the prospect of rogue worlds as planetary starships makes an exciting possibility.

However, even if a complex life could develop in the dark vents of a rogue planet's seas, it would be nothing like the view from your window. To discover more recognisable inhabitation, would it not be easier to hunt for aliens on another Earth?

PART THREE

GOLDILOCKS WORLDS

CHAPTER TWELVE

The Goldilocks Criteria

It was a time – Stephen Kane was to remark later – when hot Jupiters were still 'cool'; cool enough that a new discovery warranted a press conference. As part of the team that spotted the revealing dip in starlight from the transiting planet, Kane was describing the features of the new world to a room full of journalists. The discovery was a gas giant planet with no solid surface. This gave it a volume over a thousand times that of the Earth, which was almost entirely filled by its colossal atmosphere. At the heart of this seething gas where a solid core might exist, pressures would exceed 40 million times those on the Earth's surface. Gas at that depth would be crushed into exotic phases of metallic hydrogen that are barely producible in a laboratory. This hot Jupiter orbited so close to its star that one year took just four days. Such proximity to a stellar fireball meant that temperatures in the planet's upper atmosphere were estimated at a staggering 2,700°C (4,900°F). As Kane finished his description, one of the journalists raised their hand and asked, 'Do you think there could be life on this world?'

We are driven by an almost insatiable thirst to find habitable planets. Whether it is the excitement of meeting an extra-terrestrial being, the practical need for the human species to one day find a second home or just the concept of the unknown, the idea that there might be other liveable worlds in the Universe has dominated our fiction since the second century AD.[*]

Over the last two decades, the prospect of discovering an Earth-like world has moved from science fiction to

---

[*] The earliest-known account of alien life forms is in the novel *True Stories* by Syrian author Lucian of Samosata, who lived in 125–180 AD.

science fact. Planet detection has progressed from finding Jupiter-sized worlds packed close to their star, to planets comparable in size to our own rocky homeland. The ever-closer matches to the Earth's radius and mass has led to many of these new planets being declared *Earth 2.0*, and the prospect of a truly star-based coffee-shop franchise seems imminent. But size alone does not make our home. To understand how we might really find a habitable planet, we must consider what it means to be 'Earth'.

Possibly one of the most regrettable naming choices in planetary science is the so-called *habitable zone*. More whimsically also known as the *Goldilocks zone*, the name implies planets with crystal-clear lakes, lush greenery and a perfectly heated, oat-based breakfast. Unfortunately, the term does not mean anything of the sort.

First coined in 1959 by scientist Su-Shu Huang at the University of California, Berkeley, the habitable zone is the region around a star where water could exist on the surface of the Earth. Closer to the star and the Earth's seas would evaporate. Further away, and our water freezes into ice. The habitable zone is neither too hot, nor too cold, but just right.

Unfortunately, just as the ideal porridge temperature differs for different humans, the perfect amount of starlight differs for different planets. The starlight received in the habitable zone is unlikely to be at the right level for life-packed oceans on any planet with different surface conditions from those of the Earth. For instance, a smaller planet than the Earth might attract only a thin atmosphere and be too cold within the habitable zone to have liquid seas. At the other end of the scale, five times as many gas giants have been found within the habitable zone than rocky worlds; no one is enjoying a bowl of porridge beneath the crushing atmosphere of a Neptune. The habitable zone also does not promise that there will be water on the planet. An Earth-sized world forming from a carbon-rich protoplanetary disc would be perpetually dry, whereas a planet that was not bombarded by icy meteorites risks the same fate.

At this point, it would be reasonable to feel very annoyed. Reports in the press of new exoplanet finds frequently tout the habitable zone as a promise that the planet could support life. In fact, existence within the habitable zone does not tell us anything about what it is like on a planet's surface. It just states that if the surface were exactly like that of the Earth, your cup of water would stay liquid. To counter this nomenclatural confusion, there has been a move for scientists to refer to the habitable zone as the *temperate zone*. This emphasises an agreeable level of starlight, without directly promising bears or porridge. For the rest of this book, we adopt this convention to avoid feeling perpetually cheated.

The simplest estimate for the location of the temperate zone is found by assuming that the planet is heated just from the sunlight arriving at its location. If we do this calculation for the Earth, we would expect an average surface temperature of just 5.3°C (41.5°F). In fact, the situation is worse as the Earth reflects about a third of the Sun's heat, lowering the expected temperature to an icy –18°C (–0.4°F). This would freeze our surface water and put us out of the Sun's temperate zone, which instead would extend from 0.47 to 0.87au.* Venus could be a delightful place to live, but Earth would be a snowball. Fortunately, the Earth's true average surface temperature is 15°C (59°F), 33°C (91°F) warmer than this simple calculation suggests. The difference is that our atmosphere acts as a natural greenhouse to keep our planet warm.

Ultraviolet radiation from the Sun that passes through our atmosphere is absorbed by the planet's surface. The planet warms and reradiates this energy as infrared heat. This is easily testable on a sunny day. At high noon when the Sun is directly overhead, the ground is cool to the touch. A couple of hours later it can be too hot to walk on without shoes. This

---

* Remember, the Earth is at 1au.

delay is the time needed for the ground to absorb the ultraviolet radiation and heat to emit infrared radiation.

While the ultraviolet radiation travels through our atmosphere unimpeded, the infrared radiation cannot. The atmosphere absorbs this longer wavelength as it tries to escape, warming and reflecting part of the radiation back towards the Earth. The planet's surface therefore warms further due to the atmosphere's protective blanket. This *greenhouse effect* takes its name from the tomato-growing transparent outhouse, which heats in a similar way by trapping the infrared from the warmed air inside the glass.

How much infrared energy is trapped by the atmosphere depends on the molecules that absorb the radiation. In the Earth's air, water vapour and carbon dioxide are the two main greenhouse gases. Water vapour forms two-thirds of the infrared-absorbing molecules, while carbon dioxide makes up another quarter. The remaining few per cent are from gases such as methane, nitrogen dioxide, ozone and a splattering of human-made chloro-fluorocarbons.

If we pushed the Earth towards the Sun, the ultraviolet radiation hitting our atmosphere would increase. The temperature on the planet would rise and evaporate more water. The resulting upswing in the quantity of water vapour in the air would boost the greenhouse effect and trap heat more effectively. The surface temperature of the planet would then rise even higher.

The Earth can compensate for the rise in temperature by moderating the amount of carbon dioxide. This greenhouse gas reacts with rainwater to create carbonic acid, or *acid rain*. Upon hitting the ground, the acidic water dissolves the rocks by reacting to form carbon-rich minerals in a process known as *chemical weathering*. The dissolved minerals get swept down to the ocean and form solid carbon composites such as the chalky calcium carbonate,* and eventually limestone. This

---

* The same stuff that forms the white build-up on taps in hard-water regions, and also antacid tablets.

process removes carbon from the atmosphere and allows the planet to cool.

The carbon can be returned to the atmosphere via volcanoes. As the giant tectonic plates that form the Earth's crust collide, one sinks below the other in a process known as *subduction.* As the bottom plate forces its way beneath the upper layer, the friction between the two sections melts the rock and releases carbon dioxide. Gas and fresh silicate rock explode on to the surface through volcanoes, giving the Earth a fresh surface and expelling carbon dioxide back into the air.

This recycling of carbon is the *carbon-silicate cycle* and it acts as a thermostat to tweak the Earth's temperature. If the planet begins to warm, more water evaporates and increases the rainfall. This encourages carbon dioxide to dissolve and react with the rocks, removing it from the atmosphere. The atmosphere traps less infrared radiation with the drop in carbon dioxide and the planet cools. Conversely, if the Earth gets too cold, ice forms and the rainfall decreases. The reaction rate between the acidic water and rocks also drops in the cooler climate. Less carbon dioxide is removed from the air but is still added via volcanic activity. The amount of greenhouse gases therefore increases and traps more heat so the planet can warm.

While effective, this natural thermostat is very slow and it takes 100 million–200 million years to cycle carbon between the atmosphere, rocks and sea. It is because of this duration that the Earth's temperature rises due to human activity; we are pumping greenhouse gases into the atmosphere far faster than they can be removed via chemical weathering. While volcanoes emit a few hundred million metric tonnes of carbon dioxide per year, humans output a hundred times higher, running into the 30 billion tonne mark from the burning of fossil fuels.

Although the carbon-silicate thermostat allows the Earth to compensate for small changes in the Sun's radiation, there is a limit. If the Earth is pushed too close to the Sun, the increase in water vapour cannot be compensated for swiftly enough by the reduction in carbon dioxide. The planet therefore heats still further and more water is evaporated into the atmosphere to boost the greenhouse effect. As temperatures pass 100°C (212°F), it stops raining and carbon dioxide removal is throttled. The greenhouse gases continue to climb through water evaporation and volcanic activity, raising the temperature continuously higher. Carbon is baked out of the rocks to escape into the air and react with oxygen to add still more carbon dioxide to the atmosphere. A runaway cycle ensues whereby the planet continues to get hotter and hotter until all the water is gone from the surface.

This is probably the fate Venus met. Our neighbouring world is very close to the Earth in size, mass and composition, but boasts a thick carbon dioxide atmosphere with carbon-poor rocks and a surface temperature of 480°C (900°F). With temperatures hot enough to melt lead, no space probe has lasted more than two hours on the Venusian surface. The planet stands as a warning that Earth-sized does not mean Earth-like. Venus is definitely too hot for Goldilock's porridge.

If the Earth is now pushed away from the Sun, then the carbon–silicate cycle can boost the levels of carbon dioxide to keep the planet warm. This fails once the temperature drops sufficiently for the carbon dioxide to condense into clouds. Clouds of carbon dioxide reflect and block more of the Sun's diminishing heat to boost, rather than counter, the planet's cooling. The surface temperature on a more distant Earth would drop to zero at 1.4–1.7au. This is the point known as the *Maximum Greenhouse* limit.

The limits of the ability of the Earth's carbon–silicate cycle to mediate the temperature form the edges of the temperate zone. Closer to the Sun, the Earth would spiral into a Venus at the *Runaway Greenhouse* limit, while it would freeze into a snowball at the point of the Maximum Greenhouse limit. In our Solar System, the temperate zone extends from 0.95au to 0.14au for a conservative estimate, and 0.84au to 1.7au at the generous edge. The latter extra space allows the Earth to have water for just part of its lifetime. For instance, Mars shows evidence for once hosting water on its surface about 3.8 billion years ago. Venus may even have harboured liquid water for a brief early spell. Including these initial *recent Venus* and *early Mars* epochs expands the temperate zone to its maximum extent.

Inside the temperate zone, the carbon-silicate cycle is expected to keep the Earth's surface temperature at 0–100°C (32–212°F); the level needed to keep surface water liquid. With the limits defined by the Earth's atmosphere and geology, it is easy to see why the temperate zone is planet dependent. Boost the Earth's carbon dioxide by a factor of 10, and even our current position might be unable to support liquid water. A different mix of gases in the atmosphere or different rocks would lead to a completely different cycle from the one on our home world.

If the temperate zone only applies to one type of planet, how is it useful? The main purpose is as a target selection scheme for future astrobiology studies. A second Earth would be found within the temperate zone and there is no doubt we will recognise life most easily if it is like our own. However, being within the temperate zone does not guarantee life, water or even a solid surface.

If we keep pushing the Earth towards the Sun, eventually the intense sunlight will heat the atmosphere so strongly that the gases escape the planet entirely. As the molecules absorb the solar energy, their speed increases sufficiently to escape the gravitational pull of the planet. The point where the Sun strips the planet of its atmosphere is dubbed the *cosmic shoreline*. As is the case with the greenhouse effect, the location of the cosmic shoreline depends heavily on the planet. Lighter atoms escape more easily than heavier molecules, making hydrogen-rich atmospheres easier to strip than carbon- and oxygen-rich gases. The stripping is resisted by the gravitational pull of the planet, allowing a more massive world to withstand higher radiation. It is this stripping mechanism that was considered in Chapter 6 for turning a hot Jupiter into a chthonian super Earth. The cosmic shoreline for the Earth occurs when the Sun's radiation is 25 times stronger than our

current location, at around 0.2au. The region in between the
cosmic shoreline and the temperate zone is the *Venus zone*,
and marks the place where an Earth-like world is likely to be
overwhelmed by the greenhouse effect to become a lead-
melting Venus hellhole.

Given the unappetising appeal of Venus for bears and
porridge, it is an unfair fact that planets in the Venus zone are
easier to detect, being closer to their star than similar-sized
worlds in the temperate zone. The line between these two
regions is therefore an important point when considering the
potential habitability of exo-worlds.

🪐

A further complication to the location of the temperate zone
is the star itself. Over its lifetime, a star's brightness changes
and delivers varying amounts of heat to its surrounding
planetary system. As the star fuses hydrogen into helium
and on into the heavier elements, its core contracts. The
contraction releases energy and the star brightens. Around 3
billion to 4 billion years ago, our Sun was 30 per cent fainter
than it is today. Such a decrease in solar energy should have
meant that our planet was 20°C (68°F) cooler than at the
present time, and it was largely frozen. Rather confusingly,
geological evidence suggests that the Earth had plenty of
liquid water on its surface 4 billion years ago. Sedimentary
rocks from this epoch have been found that could only have
been created by solid particles settling out of a liquid. This
problem is known as the *faint young Sun paradox*.

The solution to this issue is still debated. One possibility is
that our atmosphere was very different billions of years ago,
containing a higher fraction of greenhouse gases capable of
holding heat. The carbon–silicate cycle may have allowed
carbon dioxide levels to rise as high as 80 per cent of the
atmosphere mass. Alternatively, early bacterial life forms may
have produced a high content of methane.

🪐

The temperate zone considers the surface temperature of an Earth-like world due to the stellar radiation. However, the star is not only a source of heat.

Along with its essential warmth, the Sun releases a continuous stream of charged particles known as the *solar wind*. This wind sweeps through the Solar System at speeds as high as 900km/s (560mps), smacking into the planets and drawing out the distinctive tails of comets. The Sun's outer layers can also erupt in localised explosions referred to as *solar flares*. These carry even more high-energy particles towards the planets. To top it all, a chunk of solar matter is sometimes hurled outwards in an event called a *coronal mass ejection*; a name taken from the outermost layer of the solar atmosphere, the corona. Coronal mass ejections can cause geomagnetic storms on Earth that interfere with our electric and GPS systems. However, the Earth is largely unaffected by the Sun's rambunctious activities as it is shielded by its magnetic field.

Travel north to Greenland or south to New Zealand, and you might be lucky enough to catch sight of the Aurora Borealis or Aurora Australis; the Northern or Southern Lights. As charged particles stream from the Sun and over the Earth, they are captured by our planet's magnetic field and channelled towards the poles. As the particles strike oxygen and nitrogen atoms in the Earth's upper atmosphere, they emit the green and blue lights of the aurora.

Without a magnetic field, the Earth would have to take the full buffering of solar particles. Our nearest neighbours are a warning that the consequences would not be pretty. Neither Venus nor Mars is protected by a magnetic field. Although both planets have very similar compositions to the Earth, small differences in their formation left these worlds without their protective magnetic shields.

Our magnetic field is generated by the Earth's molten iron outer core, still hot from the radioactive elements and collisional heat of the planet's formation. The movement of this electrically conducting metal produces a current that generates a magnetic field and turns the planet into a giant bar magnet. The motion of the molten core is controlled by our planet's rotation, and the currents that cycle heat between the core and surface. The second of these is effective because of the Earth's plate tectonics. The shuffling of the giant crustal plates exposes the hot mantle and melts old crust, releasing energy that causes our outer layer to cool. The temperature difference between core and surface creates a strong convection cycles like a giant radiator. Warm liquid rises upwards, while cooler material drops to be heated once more. This continual motion of the Earth's interior drives the molten core and our magnetic field.

While effective on Earth, neither Venus nor Mars have managed to replicate the same system. Neither planet has plate tectonics. With a surface concealed beneath thick clouds and too hot for prolonged robotic investigation, Venus's evolution is tricky to unravel. It is suspected that the absence of plate tectonics is related to the surface being too warm. The extra heat made the crust a mushy mix that allowed it to heal cracks, rather than snapping into separate plates. The lack of water on the hellishly hot surface also removed a source of lubrication, creating a far less mobile mantle. Venus additionally rotates so incredibly slowly that a Venusian day is longer than its year. It takes 243 days for Venus to rotate once, compared with the 225-day orbit around the Sun. This means that the planet actually rotates in the opposite direction from the Earth.

Slow planet rotation and the lack of strong convection currents from plate tectonics result in Venus's core rotating too slowly to generate a magnetic field. Instead, the cloud-covered world only manages to produce a very weak field in its upper atmosphere. Ultraviolet radiation from the Sun removes electrons from the very top of the Venusian atmosphere to create a layer of electrically charged particles known as the planet's *ionosphere*. Even without a core-driven magnetic field, these charges can divert the charged particles in the solar wind to produce a small current and magnetic field. This gives Venus faint aurora-like lights, but the strength is 40 times weaker than the Earth's field.

Mars's problem is the opposite to that of Venus: the planet is too cold. Part of the Martian crust is strongly magnetised, indicating that Mars must have had a magnetic field at some time in the past. This field magnetised the rocks but at some later point shut down. The problem was the Red Planet's rapid cooling. Mars's small size gives the planet a large surface area compared with its volume. Like spreading a sheet out on the washing line to dry, this wide surface allows the planet's interior heat to leak away much faster than it does on Earth. As Mars's core cooled, the convection flow between the mantle and core stopped. Any plate tectonics ground to a halt and the magnetic field died.

The magnetic field may have been given an extra death push by a major collision with a moon-sized object that struck Mars more than 4 billion years ago. At only few hundred million years old, the young planet was smacked so hard that it created a dichotomy between the two hemispheres of the Martian crust. The northern surface of the planet is an average of 5.5km (3.4mi) lower than the southern surface, with a crust that is 26km (16mi) thinner. A collision of this magnitude would have generated a huge amount of heat on the impacted northern side, resulting in a temperature gradient across the planet. This might have disrupted the convection flow through the planet's mantle and throttled the magnetic field. The soaring temperatures at the impact site would have demagnetised the rocks in that area, explaining why the magnetised rocks are found predominantly on the planet's southern surface.

Regardless of the evolutionary details, neither Venus nor Mars now has a global magnetic field. The grim results of not having a planetary shield have been observed first-hand by two space probes.

On 19 December 2006, the Sun spewed up a relatively small coronal mass ejection. This splotch of solar material travelled across the Solar System and smacked into Venus four days later. The impact was watched by the European Space Agency's *Venus Express* probe, which was orbiting the planet to examine its atmosphere. Despite being a small and slow puff from the Sun, the coronal mass ejection ripped away a substantial amount of oxygen from Venus's unshielded atmosphere. *Venus Express* also spotted hydrogen and oxygen being blown away from the unprotected Venusian air by the solar wind; removal of the last vestiges of Venus's seas.

The effect such events have on Mars was later spotted by the NASA probe, *MAVEN*, which was doing a similar job to the *Venus Express* around our outer neighbour. On 8 March 2015, *MAVEN* viewed the impact of a much stronger coronal mass ejection hitting Mars. The atmosphere loss from the Red Planet was bumped by a factor of 10. The continuous solar wind also depletes the Martian gases, ripping away about 100g (3½oz) each second from the small world.

These problems on Mars and Venus would have been substantially worse in the past. A young star is an excitable entity and sheds far more material into space than our current, mellow Sun. The fact that Mars appears to have hosted liquid water in the past suggests that its atmosphere was once thick enough to warm the planet. With the loss of its magnetic shield, the Sun helped to rip away Mars's gas blanket to leave the present uninhabitable world.

The surface habitability of any exoplanet probably depends on the planet having a magnetic shield. While no observation can presently detect a magnetic field, it is a factor to remember before declaring a new planet find 'Earth-like'.

We have come a long way from the first discoveries of the hot Jupiters in the 1990s. Worlds are now being found that are close to Earth in size and orbit within the temperate zone. But are these really likely to be Earth-like enough to be Earth 2.0?

CHAPTER THIRTEEN

The Search for Another Earth

Just three days after the start of its scientific operation, the Kepler Space Telescope got lucky. It spotted the transit of a planet inside the temperate zone of its star.

It would take another two-and-a-half years to certify the find. The characteristic dip in starlight needs to be seen at least three times to confirm the presence of a planet and measure its properties. The telescope had caught the first transit almost immediately, in May 2009. Another two had followed by December 2010. One year later, on 5 December 2011, the announcement was officially made: the first transiting planet within the temperate zone had been discovered. 'Fortune,' said William Borucki, who had led the discovery team from the NASA Ames Research Center in California, 'smiled upon us with the detection of this planet.'

The new world was Kepler-22b. The planet orbits a Sun-like star 600 light years away in the constellation of Cygnus, the Swan. Sitting at a distance of 0.85au, Kepler-22b loops around its star every 290 days. Around our own Sun, this location would place Kepler-22b right on the edge of the early Venus boundary for the temperate zone. If the planet was also Earth-like, it would suggest that the world could have harboured liquid water for a brief period in its early history. However, Kepler-22 is a slightly smaller and cooler star than our Sun, making it 25 per cent less luminous. The weaker radiation pulls the temperate zone inwards to place Kepler-22b firmly within its conservative boundary. Did this make the planet our first look at another Earth?

The media were certain: 'Earthlike planet found orbiting at right distance for life', blared the headline on the *National Geographic* website. 'Kepler-22b – the "new Earth"', screamed the *Telegraph*. 'Earth-like planet confirmed', added the BBC.

However, as the planet's measurements rolled in, it became clear that this world was keeping its secrets. The planet's radius is 2.4 Earth radii, placing Kepler-22b in the mysterious super Earth category between the sizes of our rocky and gaseous worlds. It is too small and distant to produce a detectable wobble in the star's position, making it impossible to measure the planet mass. This ruled out the option of calculating a bulk density to reveal if the planet was probably terrestrial, or a lot of hot gas.

Radial velocity data would also have allowed the eccentricity of the planet's orbit to be estimated. A transit observation sees only one part of the orbit as the planet passes across the star's surface, while the radial stellar wobble maps the planet's full circle. With only the transit detected, the planet could be on a bent path that spends just a tiny fraction of its year inside the temperate zone.

Based on the rule of thumb discussed in Chapter 6, it seems unlikely that a planet larger than 1.5 Earth radii would be rocky. However, Kepler-22b's intermediate size might make it a water world. Such a planet would have a rocky core entirely engulfed by an ocean thousands of kilometres thick. With the sought-after temperate zone based on the concept of surface water, a global sea might seem a positive feature for life. The problem is that the lack of land throttles the carbon-silicate cycle, as we will see in the next chapter. Life is not necessarily precluded on such a world, but it would certainly be different from that on Earth. Kepler-22b may have been a first discovery, but it is not a second Earth.

In 2010, the most exciting place in the Galaxy was the neighbourhood of the red dwarf star Gliese 581. The small star has only a third of our Sun's mass and sits 20 light years away in the constellation of Libra, the Weighing Scales. Around its moderate light were thought to be six planets, all between the mass of the Earth and Neptune. It was our Solar

System in miniature, and intriguingly, three of those worlds looked potentially habitable.

Red dwarf stars with masses of between a tenth to half that of our Sun are exciting prospects for detecting small planets. For a start, these dim stellar furnaces are numerous and form about three-quarters of the stars in our Galactic neighbourhood. Their small size decreases the ratio between the planet and star, making both the transit light dip and radial velocity wobble more pronounced and easier to detect. Finally, their low luminosity places the temperate zone much closer to the star. The proximity boosts the chances of a planet within the temperate zone transiting across the stellar surface, since the orbit would have to be highly inclined to miss the star's face entirely. The short year in the closer orbit also leads to frequent transits, giving multiple chances to spot the planet. The net result is that rocky worlds in the temperate zone are easiest to find around red dwarfs.

Between 2005 and 2010, six planets were discovered orbiting Gliese 581 using the radial velocity technique. The first to be announced was predictably the weightiest and orbited close to the star; Gliese 581b was a Neptune-sized world at almost 16 Earth masses, with an orbit time of a little over five days. Next to be discovered were two super Earths, Gliese 581c and Gliese 581d. This pair had masses of 5.5–6 Earths, and orbited in 13 and 67 days. Then a planet only twice the mass of the Earth was found; Gliese 581e circled the star closer than its three planetary siblings in just 3.1 days. Finally, two more distant super Earths were detected. Gliese 581f was 7 Earth masses and had an orbital time of 433 days, while Gliese 581g was 4 Earth masses with an orbit that lasted just over a month.

With the exception of the outermost super Earth, Gliese 581f, all the planets orbited far closer to their star than any planet in our Solar System. But the red dwarf's dim light indicated that these were not scorched worlds. Instead of a temperate zone around 1au, the location where liquid water could persist on an Earth-like planet lay at 0.09–0.23au, corresponding to circular orbits lasting 18–72 days. This

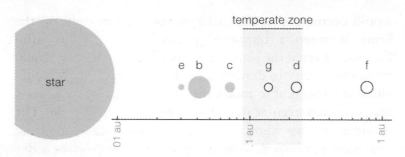

Figure 21 Gliese 581 is a red dwarf star that was thought to be orbited by six planets, two inside the temperate zone. However, later observations questioned the existence of planets d, f and g.

placed planets Gliese 581d and Gliese 581g inside the temperate zone, while Gliese 581c sat tantalisingly just beyond the inner edge. Could one of these three worlds be Earth-like enough to be awash with an ocean?

Upon its discovery in 2007, Gliese 581c was the lowest-mass exoplanet ever found. Despite orbiting slightly inside the inner edge of the temperate zone, there was an optimistic proposal that a covering of reflective clouds might be able to keep the planet cool. Reflecting 50 per cent of the star's radiation at the location of Gliese 581c would allow an otherwise Earth-like planet to have an average surface temperature of 20°C (68°F). While the Earth reflects only about 30 per cent of sunlight, Venus's clouds bounce back 64 per cent of the incident rays. Fifty per cent reflection therefore seemed possible, and Gliese 581c was declared in its discovery paper the 'most Earth-like of all known exoplanets'. It was a bold proclamation, but could a few clouds truly make Gliese 581c a serious contender for a habitable world? Unfortunately, the odds are stacked against a cooler climate.

The first problem is the planet's location. Even adjusting for the red dwarf's weak heat, Gliese 581c sits closer to its star than Venus does to our Sun. Moreover, while Venus's clouds may be reflective, they also indicate the suffocating atmosphere of a runaway greenhouse environment.

This risk is compounded by the planet's mass. If Gliese 581c has an Earth-like composition, its 5.5 Earth masses would correspond with 1.5 Earth radii; a size right on the brink between a terrestrial planet and a gaseous mini Neptune. Even if the planet were rocky, the extra mass would boost the gravity to draw in a thick atmosphere. This would efficiently trap heat to rocket the surface temperature beyond that expected even at the edge of the temperate zone. The stronger gravity also risks the primitive atmosphere of hydrogen and helium gases being retained to produce a dry and unusable mess.

Just in case a brief holiday still sounded tempting, the proximity of Gliese 581c to its star indicates that the planet risks tidal lock. Like the CoRoT-7b lava world, a tidally locked Gliese 581c would be a split world of day and night with one face permanently turned towards the star's inferno. Such divided worlds can struggle to distribute heat around the globe. This does not necessarily render a world barren, but such a temperature dichotomy is unlikely to aid life. The combination of these factors is enough to take Gliese 581c out of serious consideration for habitability.

In contrast to this, the main issue with the habitability of planets Gliese 581d and Gliese 581g is that they may not exist. Two weeks after the discovery of Gliese 581f and Gliese 581g was announced, their existence was called into question at a meeting of the International Astronomical Union in Italy. Fresh observations had confirmed the presence of planets b, c, d and e, but failed to find definite signatures for planets f and g. The difficulty was that teasing apart the wobbling motion of the star to pick out the rhythmic tugs from multiple orbiting planets is immensely tricky. This is especially true for red dwarfs, which are intrinsically faint and typically rambunctious stars. Even slight fluctuations on the star's fiery surface can lead to a false find.

The non-existence of Gliese 581f was accepted, but researchers haggled over Gliese 581g. Further analysis failed to reach any conclusion: was this planet real or just a ghost? If the planet existed, it sat squarely in the temperate zone. Moreover, with a mass just three times that of Earth, Gliese 581g was far more likely to be rocky than Gliese 581c. With the chance of habitability being waved like a carrot on a stick, everyone wanted this planet to exist.

In 2014, these hopes were dashed. Further observations of Gliese 581 had recorded unusual magnetic activity on the star's surface. The magnetised patch was similar to a sunspot and was interfering with the surrounding flow of stellar material. As the star rotated, the spot appeared as a periodic wobble that looked strongly like the influence of a planet. When this effect was removed from the data, Gliese 581g vanished. What was worse, the correction also erased Gliese 581d. This second casualty had been measured to have an orbit taking twice the time of Gliese 581g and turned out to be linked with the same anomaly.

While the observations of Gliese 581 remain debated, the prospect of not existing is a blow to the real estate of Gliese 581d and g. Hunting low-mass planets was proving to be an extremely difficult game.

There are two problems with finding a transiting Earth-sized planet in the temperate zone. The first is that a planet and star analogue to our own has only a 0.1 per cent chance of transiting. From most viewing angles, the small and distant Earth does not cross the Sun's face. The second issue is that the decrease in light as the planet crosses the star is just one part in 10,000. 'Imagine the tallest hotel in New York City, and everybody has their light on,' Natalie Batalha, project scientist for the Kepler Space Telescope, stated. 'And one person in his hotel lowers the blinds by 2cm. That's the change in brightness we're trying to detect from the transit of a planet as small as Earth passing a star the size of our Sun.'

Yet on 18 April 2014, the Kepler Space Telescope science team announced that it had done it. Kepler-186 was a red dwarf about 500 light years away in the constellation of Cygnus. With half the mass of our Sun, the star's dim light pulled the temperate zone inwards to sit at 0.22–0.4au, almost entirely within Mercury's 0.4au orbit in our Solar System.

The planet was Kepler-186f, which sat on the outer edge of the conservative temperate zone with an orbital time of 130 days. It had a radius of 1.11 Earths, tantalisingly close dimensions to our own planet. At such a small size, Kepler-186f was unlikely to be anything other than rocky.

Like the Earth, Kepler-186f is part of a system of planets. Four other worlds had previously been discovered, all with sizes smaller than 1.5 Earth radii. These four siblings orbited closer to the star than Kepler-186f, taking 4–22 days to loop the red dwarf. While also small enough to be rocky, the planets sat inside the inner edge of the temperate zone, and were liable to be too hot to support liquid water even if they mirrored the surface conditions on Earth. Unlike in our own system, the most likely candidate for habitability was the outermost planet. Could we now finally say we had found an Earth twin?

The only true way to ascertain whether Kepler-186f is Earth-like would be by exploring its surface. While we cannot yet send spacecraft between star systems, clues could be gleaned from the planet's atmosphere. As we saw for 55 Cancri e, the light passing through the gases surrounding a planet transiting across its star can reveal information about surface conditions. For instance, the Earth's atmosphere is packed with oxygen and methane from the teeming life on our ground. Unfortunately, Kepler-186f is 500 light years away, rendering it too distant and small for an atmospheric study. The best we can do is speculate.

In fact, the position of Kepler-186f raises a couple of interesting problems. The first is a ubiquitous issue with the temperate zone around red dwarfs. Situated close to the star, planets in this potentially comfortable region travel on much shorter orbits than the Earth. Forming at such a location would be a rapid process. Able to whip around the star almost three times as often as the Earth, collisions between planetesimals would be frequent and material would quickly accumulate. This initially sounds very positive; swift planet formation would allow a longer time in the temperate zone to develop a life-supporting environment. But there is a price.

The very young red dwarf is a hot beast. Before the start of nuclear fusion, proto-red dwarf stars are surprisingly luminous. Unlike larger Sun-sized stars, a forming red dwarf can be 100 times brighter than its normal dim value once it begins to burn hydrogen into helium. If a planet has formed during this early phase, any surface water could be evaporated away before the star cools. The surface temperature on Kepler-186f might allow liquid water now, but possibly there is none left on the planet to form a sea.

An associated problem with forming close to the star is the speed of the planetesimals and embryos. These rocky bodies move rapidly on close-in orbits. The final planet-formation stage may end up being dominated by high-velocity collisions, capable of stripping away atmosphere and water from a young world.

A second issue is that Kepler-186f appears to be the outermost planet around its star. For the Earth, this is of course not true. Beyond our orbit sits Mars, then the neighbourhood of the gas giants. The presence of Jupiter has been particularly important in our evolution, since its powerful gravity is thought to have scattered ice-rich planetesimals inwards to deliver our oceans. Admittedly, the gas giant games of gravitational pinball also present risks to a young planet, but a lack of water would definitely preclude Earth-like life. Without such an outer kickballer, would Kepler-186f be a dry world?

These seem like significant problems for Kepler-186f. Yet, the layout of the planetary system could indicate an alternative and more promising history. This system has five worlds that all sit very close to the star. For these to form on their current orbits, the original protoplanetary disc would have needed to contain more than 10 times the Earth's mass, with most of this bulk within 0.4au of the star. Such a shape is not commonly observed in discs around young stars. It is therefore more likely that the Kepler-186 planets formed further out and migrated inwards. This circumnavigates the above concerns: forming in the colder outer reaches of the disc would allow ice to solidify with the planet and form water-rich worlds.

These could then migrate inwards from the gas drag after the star had outgrown its fiery protostar phase. The outer location of Kepler-186f would then prove an advantage, since it may be distant enough to avoid tidal lock. With a regular rotation that allows the planet to be evenly heated, surface water could possibly be maintained.

And yet... migration does not entirely bypass all problems. Kepler-186f is still close enough to the star to feel the full frontal of space weather from the star's stellar wind. Without a strong magnetic field, the planet could find itself stripped of its air. Whether the planet is able to generate a magnetic field will be down to its geology. But while Kepler-186f is a likely size to be rocky, there is no way of telling what types of rock it contains.

As the varied scenarios for the composition of 55 Cancri e testify, rocky does not necessarily mean Earth-y. Even a mix of iron, silicate and ice can lead to very different planet masses. A pure iron Earth-sized world would have a mass of nearly 4 Earths, while one dominated by ice might only weigh in at 0.32 Earths. If Kepler-186f did have the same mix of iron and silicates as our planet, its mass would be 1.44 Earths. So while the planet's size might be only 10 per cent larger than the Earth, its mass could be between a third to one-and-a-half times our bulk. These variations would produce strong differences in the planet's gravity and internal pressure. The resulting rock composites may not have the right mobility to generate a magnetic field. The difference in gravity will also affect the atmosphere gases that can be drawn in and held by the planet.

But yet again, we can argue this from the opposite direction. Models of the effect of stellar flares and wind on a planet without a magnetic field around a red dwarf have suggested that the damage may be restricted to just the upper atmosphere. This could leave conditions on the ground unharmed by the rambunctious star. Until we probe the atmospheres of more small planets beyond our Solar System, the effect of non-Earth environments involves a lot of guessing.

It is also worthy of note that life on Earth can be found in the most unappetising places. So-called *extremophiles* are creatures that can survive (as their name suggests) in extreme levels of hot and cold temperatures, acidity, pressure and dryness. One of the most resilient examples are the *tardigrades*, or *water bears*; eight-legged micro-animals that can successfully hibernate in temperatures of $-256–+151°C$ ($-428–+304°F$), pressures higher than those found in the ocean trenches, and exposure to radiation levels hundreds of times higher than would kill a human. However, whether life could begin in such extreme conditions or can simply evolve to adapt is a big unknown.

As for Kepler-186f, it is possible that it may be habitable and host life. We can say that its location and size do not rule this out, but we equally cannot say that these factors guarantee hospitable conditions. Orbiting a red dwarf, any life on Kepler-186f would be very different from our own. At high noon, the close star would look a third larger in the sky than our Sun, but its brightness would match the Sun on Earth an hour before sunset. This dimly lit distant world could perhaps be a distant Earth cousin, but it can never be a twin.

By November 2016, 93 planets had been confirmed orbiting entirely within the temperate zone of their star, with 217 planets found with at least part of their orbit inside this region. Of these, five planets had radii below 1.5 Earths and were likely to be rocky, with Kepler-186f being the smallest and closest in size to the Earth.

What does this tell us about the rarity of worlds that could be potentially Earth-like? While the number of small planets found is low, the total number of new worlds discovered is huge. It is huge enough to do some statistics.

Based on the 2,300 planets that had been discovered by the Kepler Space Telescope by 2013, it was estimated that one in six stars had a planet within 80–125 per cent of the size of the Earth. Around the Milky Way's 100 billion stars, this would mean that 17 billion Earth-sized worlds are out there. The calculation used for this excitingly big number had included an estimate for the number of planets that might have been missed in observations and the chances of a false detection. However, it only applied to planet orbits that were shorter than 85 days. For longer orbits, the number of discovered planets was still too low for a meaningful calculation.[*] With such short years, most of these 17 billion worlds would be too hot to be inside the temperate zone.

To combat this problem, a second estimate was made for planets orbiting red dwarfs. Smaller worlds around these stars are easier to observe, especially in the temperate zone where the closely orbiting planet would transit about five times in one Earth year. From an examination of nearly 4,000 dwarf stars, just under 40 per cent have a planet that is likely to be rocky; 15 per cent of these were also within the temperate zone of the star. This implied that there was most probably an Earth-sized planet in the temperate zone within 10 light years of Earth. It was an intriguing thought. Where was our nearest rocky planet?

In the summer of 2016, it looked as though we might have the answer. A planet had been discovered around Proxima Centauri, the dim third wheel that formed the triple star system with the Alpha Centauri binary.

Of the three stars, Proxima Centauri sits nearest to the Earth. Its distance is 4.22 light years, compared with the 4.3 light years to Alpha Centauri. The distance between the binary and third star is a rather considerable 13,000au and casts doubt over whether the trio are truly held together, or if Proxima Centauri is just passing through the system. Whichever turns out to be the case, Proxima Centauri is our nearest neighbour, making any planet the star hosts the closest

---

[*] Note that this does not mean that there are fewer planets further from their star, only that these are trickier to detect.

possible exoplanet to us. The discovery of Proxima Centauri-b was therefore justifiably exciting.

The planet had been found using the radial velocity technique, providing a minimum mass of 1.3 Earths. With no observed transit, the orientation of the planet's orbit remained unknown, leaving the exact mass a mystery. If we were seeing the orbit of Proxima Centauri-b at an angle larger than 15 degrees, the planet mass would be in the mini Neptune regime. That said, it is still more likely that our nearest neighbour has a mass commensurate with a rocky planet.

The planet orbits Proxima Centauri at just 0.05au, giving it a year of 11.2 days. This might suggest that the planet is a baked lava world, but Proxima Centauri is a dim star even for a red dwarf. With just over 10 per cent of the Sun's mass, the star's radiation is so weak that Proxima Centauri-b sits within its temperate zone.

Of course, the present weak flow of energy from the star does not avoid the problems faced by Kepler-186f in orbiting a red dwarf. Proxima Centauri is a particularly active star even today, suffering from mega flares that periodically bathe its closely orbiting planet in radiation levels hundreds of times higher than those found on Earth. Unless Proxima Centauri-b is protected by a strong magnetic field, its atmosphere may well be stripped.

The loss of atmosphere would be particularly bad since the planet's incredibly close orbit indicates that it is almost certainly tidally locked. Without an atmosphere capable of redistributing heat, the planet will be split into roasting and freezing hemispheres, corresponding to perpetual day and night.

The star's strong activity also comes with the risk that this planet may be a false detection. With a lot of action and change on the star's surface, picking out the tiny wobble from an exoplanet becomes an even greater challenge.

In spite of these concerns, the proximity of Proxima Centauri-b makes this find one of the most exciting exoplanet discoveries. If future observations are able to examine the planet's atmosphere, we may get a peek at what surface conditions are like around red dwarfs. The easiest way to achieve this would be if the planet were found to transit the star. Unseen so far, the chances are not high but Proxima Centauri is being watched carefully for signs of a periodic dimming. A second option would be a direct image of the planet. Direct imaging is always a challenging prospect, especially for such a small world. However, Proxima Centauri-b is our nearest possible exoplanet. As new space telescopes such as the Hubble Space Telescope's successor the *James Webb Space Telescope* (JWST) and the *Wide Field Infrared Survey Telescope* (WFIRST) come on line, together with the ground-based *Extremely Large Telescope* (ELT) and *Thirty Meter Telescope* (TMT),* this planet will be one to watch.

As our nearest exoplanet, how long would it take to visit Proxima Centauri-b? While 4 light years sounds tiny compared with the 500 light-year distance of Kepler-186f, a light year is a blisteringly long distance. The furthest humans have ever travelled is a loop around the Moon; a teeny 0.00000004 light-year distance. Our furthest and fastest travelling spacecraft is *Voyager 1*, which would still take 75,000 years to reach Proxima Centauri if it were orientated in the right direction.

Other ideas have been floated for attempting high-speed travel of miniature probes, but these are no more than drawing-board sketches at this stage. For the moment, the district of our nearest stars must be studied through a telescope's eye.

* Let's take a moment to appreciate the descriptive, yet unimaginative, naming of astronomical instruments.

CHAPTER FOURTEEN

Alien Vistas

O f the thousands of discovered worlds, there is only one that can definitely support life: the Earth. This has focused the search for habitable planets on worlds that might be like our own.

It is true that an inhabited Earth-like environment will be the easiest to recognise. Yet, it is not necessarily true that this is the only option for supporting life. The Earth may not even be the best location. So what alien landscapes might still support Goldilocks's perfect breakfast?

Water worlds

The first transiting planet found inside the temperate zone raised hopes of a life-packed land. Kepler-22b seemed too large to be rocky, but too small to be a gas giant. Could it be an intermediate liquid world, blanketed by a deep global ocean? Since we find life wherever there is water on the Earth, this prospect had the splash of potential.

The first question is whether such a water world could truly exist. Without a mass measurement to provide an average density, Kepler-22b was a wild hope. Size alone was not enough to distinguish a new type of ocean planet from a small Neptune or giant terrestrial world. However, stronger evidence for water worlds had already been found.

In 2009, a planet was discovered transiting a red dwarf 42 light years away in the constellation of Ophiuchus, the Serpent Bearer. Even without allowing for an atmosphere, the planet's surface temperature was estimated at well over 100°C (212 °F), and definitely not within the temperate zone. However, the planet's close 1.6-day orbit meant that its presence wobbled the star enough to secure a radial velocity measurement of its mass. Combined with the transit data, this gave an average density.

The planet was Gliese 1214b, and was found to have a radius of 2.7 Earths and a mass of 6.6 Earths. The resulting density was $1.87g/cm^3$, which sits in between the rocky Earth and gaseous Neptune. A possibility that matches this middling value is a composition of 25 per cent rock and 75 per cent water, surrounded by a hydrogen and helium atmosphere. For comparison, our planet has only a teeny 0.1 per cent of its mass in water. The high temperatures on the planet would prevent this huge water reservoir from forming a liquid ocean. Instead, the world would be enveloped in the liquid-like gas of a supercritical fluid.*

The Hubble Space Telescope attempted to put the watery nature of Gliese 1214b to the test by examining its atmosphere during transit. Unfortunately, it failed. Rather that finding the distinctive fingerprint from water molecules absorbing the starlight, there were no distinctive features at all. The most likely explanation for such a nondescript result is that clouds are obscuring the telescope's view. Despite this lack of final confirmation, the density of Gliese 1214b makes it a prime candidate for a world made predominately of water. Its discovery secured water worlds as science (very nearly) fact.

To gather so much water, Gliese 1214b would have had to form far from the star. Behind the ice line within the protoplanetary disc, the planet could accumulate its large water fraction in frozen ices. The pull from the gas disc then caused the ice-rich world to migrate closer to the star. If its final stopping place had been in the temperate zone, a liquid ocean water world would have been born.

Water worlds may even not be rare. Gliese 1214b and Kepler-22b are both super Earths; the most common class of exoplanet currently discovered. With sizes a few times larger than the Earth, this planetary group could include two very different types of water world.

---

* We met this bizarre state of matter in Chapter 7, as a possibility for the composition of 55 Cancri e.

Gliese 1214b is an example of a deep-ocean water world. Its middling density implies that a huge fraction of its mass is in water. The planet's rocky core would be buried beneath oceans tens of thousands of kilometres deep. But the planet does not need to be mainly water to have a global ocean. The stronger gravity on a super-sized rocky planet could flatten the surface topology. Instead of mountainous terrain, the landscape could be compressed to form a featureless ocean bed that could easily become covered with a shallow sea. The difference between these water worlds is like pouring water on to a bowl and a plate. They might both be covered, but one contains far more liquid than the other. A 10 Earth mass rocky planet risks not having exposed continents unless it is very dry, with at least 10 times less water than our own planet. A larger version of our own Earth may therefore always be a water world.

Life on Earth is found wherever there is water. Yet, could an ecosystem really develop on a world without any dry land? If a global ocean renders a planet uninhabitable, this places a size constraint on even rocky planets for developing life.

On a deep-ocean water world, any plants that use light for photosynthesis would need to manage without an anchor. Stems and roots could not bridge the depth of more than 10,000km (6,000mi) between the ocean floor and the starlit surface. Floating plants like algae would need to develop, alongside creatures that could fly or swim. Of course, the Earth offers examples of such life as well as hosting organisms that do not need sunlight at all. Could the ocean-based life on our planet thrive on a deep-ocean water world?

Despite only 0.1 per cent of the Earth's mass being in water, the bottoms of our oceans are too deep for light to penetrate. Instead, hydrothermal vents spew hot fluid through cracks in the ocean crust. Despite reaching temperatures of well over 100°C (212°F), the towering geysers remain liquid due to the high pressure at the ocean bottom. Hydrothermal vents are surrounded by whole ecosystems that thrive entirely without sunlight. It is this source of energy that might support life on a rogue world, where there is no star to provide light anywhere on the planet. Unfortunately, attempting to apply this system to a deep-ocean water world sends us to a screeching halt.

Switch the Earth's 0.1 per cent water mass for more than 50 per cent and the bottom of the ocean becomes a very different place. Beneath such a huge reservoir, the pressure becomes high enough to compress the water into thick layers of ice. The silicate rocky core is thus separated from the watery ocean by an icy barrier thousands of kilometres thick. Trapped below the ice, hydrothermal vents cannot form and the potential ecosystem is silenced.

The lack of exposed land also shuts down the carbon-silicate cycle. This is the planet thermostat we met in Chapter 12, which adjusts the surface temperature by varying the amount of carbon dioxide in the atmosphere. Should the planet's temperature rise, more carbon dioxide is drawn out of the atmosphere by reactions with surface rocks. If the planet cools, this reaction slows and carbon dioxide levels increase to trap more heat. The absence of exposed rocks throttles this planetary temperature controller.

Could this be fixed by reactions with the ocean itself? The Earth's seas have absorbed 10 times more carbon dioxide than is present in the air. This would also happen on an ocean world, but it turns out that the mechanism is the Devil's thermostat. Sea absorption of carbon dioxide is most efficient when the temperature is cooler. If the planet's temperature rises, the sea will draw less carbon dioxide from the atmosphere and allow more heat to be trapped. Should the reverse happen and the planet cools, the sea will increase the removal of carbon dioxide and let more heat escape. Rather than countering a change in the planet's temperature, the endless oceans will accelerate it.

Without the ability to compensate for variations in temperature, the temperate zone for a deep-ocean water world narrows to a thin strip. Removal of the Earth's ability to adjust for slightly too much or too little radiation from the star allows the super Earth to retain liquid oceans only if its location is perfect. This does not make it impossible for the planet to be habitable, but is does make the chances of it being in a suitable place much smaller. A silver lining is that a huge ocean changes temperature very slowly. A deep-ocean world on an eccentric orbit that passes through the temperate zone might therefore be able to retain habitable conditions where changing temperatures on an Earth-like world would sterilise the surface.

Should the planet be a shallow-water world then prospects slightly improve. Without the colossally deep oceans, high-pressure ices will not form and separate the rocky ocean bed from the seas. This allows both hydrothermal vents to form and also a weak carbon-silicate cycle. Seawater can recycle carbon dioxide into the rocks in the same process as on the surface. The catch is that the ocean floor does not perfectly reflect the surface temperature of the planet, producing a much poorer thermostat. A shallow-water world may be less able to maintain its environment compared with the Earth, but will probably do better than its deep-ocean counterpart.

While the situation still looks bleak for habitability, an important caveat might rescue shallow-water worlds. The surface of a terrestrial planet is not the only place where the world can store water. The Earth tucks a significant fraction of its seas into the mantle, absorbing the water in the rocky minerals. The surface and mantle reservoirs exchange water during the shuffling of the tectonic plates. As the ocean crust is pushed underground, water is expelled into the mantle. This is then returned to the surface via volcanoes. The stronger gravity on a super Earth will increase the pressure between the two reservoirs and allow more liquid to be pushed into the mantle. If enough can go underground, even the flat sea bed of the super Earth may become partially exposed. With land out of the global bath, we would lose the water world but regain a carbon-silicate cycle.

It is difficult to determine how effective stuffing water below ground would be. This is partly because we do not know how much water is stored in the Earth's mantle. If it is approximately the same quantity as that of the surface oceans, then a 10 Earth-mass planet with plate tectonics could avoid a water world's fate. This is a reasonable or even slightly conservative guess, suggesting that larger rocky planets are not out of the habitability stakes.

It is interesting to speculate whether intelligent life could develop on a habitable water world. Without dry land, fire and electricity might never develop and this could prevent the creation of a technologically advanced civilisation. The stronger gravity on a super Earth might also limit the size of any flying life forms. Our own oceans are teeming with massive sea life, but none has reached the intelligence of humans. Is this random chance, or could evolution in the seas not favour cognitive developments?

A gas giant core

Unlike in our own Solar System, gas giant planets do not always keep to the far side of the ice line. Instead, the pull of gas within the protoplanetary disc can cause the young giants to migrate towards the star. Should migration move the planet into the temperate zone, we are left with the obvious question: is there any hope for life?

Our own gas giants do not look promising. It is impossible for a liquid ocean to exist under the colossal atmospheres that roll across these huge planets. The pressure near the solid cores of such worlds is so high that strange forms of matter are thought to exist, including liquid diamond and metallic hydrogen. Any organism attempting to live suspended within the gas would be at the mercy of strong convection currents and continuously slammed between scorching-hot depths and the freezing upper atmosphere. It is certainly an intriguing environment, but not one for a holiday resort.

The story might be different if the migrating planet is small enough to be a mini Neptune. Dragged towards the star, the increase in radiation might be sufficient to strip this intermediate-sized planet of its thick atmosphere. The result would be an exposed solid core. Could such a surface be habitable?

Planets around red dwarf stars are particularly promising for this possibility. The star's dim light allows the inner disc edge and ice line to sit much closer to the temperate zone boundaries than around a Sun-like star. This increases the chances of a migrating gas planet finding itself stranded in the temperate zone region once the gas disc disperses. Once there, the planet needs to lose its atmosphere and this is where a red dwarf may excel.

Previously, the violence of a young red dwarf was flagged as a problem for life. Planets forming in the temperate zone risk being stripped and sterilised from the strong radiation pouring from the rambunctious protostar. However, a migrating planet with a thick atmosphere can use this radiation to uncover its core.

Such a planet would form beyond the ice line and develop a centre of rock and ice surrounded by a deep hydrogen and helium atmosphere. While the red dwarf was still young, the planet would migrate into the temperate zone. The X-rays and ultraviolet radiation pouring from the protostar would heat the planet's atmosphere, allowing the gases to escape the gravitational tug of the planet. The ice on the exposed core would melt to leave a world with an ocean.

Whether this mechanism is successful comes down to size and timing. If the core is too massive, then its gravity will resist the radiation bombardment and hold on to the hydrogen and helium atmosphere. A core around the mass of the Earth can be successfully stripped of its inhospitable envelope, while one twice the mass will probably retain the gases. Additionally, if the planet moves into the temperate zone too early, then the young star's violent days may last long enough to rip away the atmosphere and still evaporate all the water. Conversely, arriving too late risks missing the star's energetic youth to leave radiation too weak to deal with the gas. But should the planet arrive with the right mass and the right timing, an exposed watery core may result.

An exposed core would offer a very different landscape from a terrestrial planet's surface. Rather than consisting primarily of rocky silicates, a core would have a comet-like composition with equal parts of ice and rock. If too much of the ice melts, the planet is likely to become a water world.

Having shed its first atmosphere of hydrogen and helium, a habitable exposed core would need a new envelope of gases. The Earth formed its second atmosphere through volcanoes releasing gases trapped in the planet's interior. Due to the cometary composition of a core, a similar expulsion would eject air rich in ammonia and methane; both greenhouse gases efficient at trapping heat at a planet's surface. The ideal location for an exposed core might therefore be at the outer edge of the temperate zone, where a boost in surface temperature would not send the planet spiralling into a runaway greenhouse.

The exposed core's rock and ice mix would also change the tectonics and geology from that on Earth. The result of this is not known, but the core's new atmosphere would be more resilient to any stellar activity if the planet can still drive a magnetic field.

As planetary migration appears to be common in exoplanet systems, it is worth remembering that rocky planets in the temperate zone may be exposed cores. If these could indeed be inhabited, their life will have developed in an extremely alien land.

The twilight zone

Along with the perils of sterilising radiation, planets in the temperate zone around red dwarf stars risk tidal locking. Orbiting so close to the star, the strong gravitational pull forces one side of the planet to continually face inwards, while the other (quite literally) never sees the light of day.

A guide to how challenging this might be for habitability is to consider what would happen if the Earth became tidally locked to the Sun. We saw in Chapter 12 that without any atmosphere, the Earth's average temperature would be about 5°C (41°F). In tidal lock, the Sun pounding on one half of our planet would send the day-side temperature soaring to 120°C (248°F). Facing away from the Sun, the night side would be warmed only by the Earth's internal heat. This paltry energy source would give a surface temperature of -273°C (-459°F). Our present Goldilocks conditions would be replaced by the unappetising options of dying by boiling or freezing.

But this bleak prospect ignores the planet's atmosphere. While the surface of the planet is locked in place, the enveloping gases can still move around the globe. Is this circulation enough to quench the extremes of day and night into liveable conditions?

The situation does not initially look good. The night side is so cold that gas would condense on to the surface. The loss of atmosphere over the dark hemisphere would cause the pressure to plummet, sending gas from the day side pouring around the globe to fill the void. As fresh gas hit the freezing temperatures, it would also condense until the whole atmosphere was destroyed. This planet would have suffered complete *atmospheric collapse.*

The catastrophic atmospheric collapse could be avoided if the atmosphere can even out the heat between the hemispheres. If the night side can be kept warm enough to prevent gas from freezing, then the atmosphere stays gaseous. A very thin envelope will not be able to move enough hot gas around the planet to prevent freeze-out, but an approximately Earth-like carbon dioxide or nitrogen atmosphere may succeed.

Whether an Earth-like pervading atmosphere would allow lakes and seas to form is another matter. Success depends on whether the planet's day side becomes hot enough to boil water. On the night side, water is doomed to be solid ice. With temperatures low enough to risk freezing the atmosphere, not even circulating air will raise conditions sufficiently to keep water liquid in the dark. If heat from the star or a thick greenhouse atmosphere evaporates the water on the day side, the steam will be blown to the night side of the planet as the atmosphere circulates. Now below the freezing point for water, the vapour will condense into snow and fall to the icy surface. The night side will become a cold trap that will eventually hold the planet's whole water budget as frozen ice. Such a planet would look like a giant eyeball, with a surface covered by ice except where the planet faces the star.

Even at temperatures of less than 100°C (212°F), eyeball worlds are a risk. Any water that evaporates from winds blowing across the seas will be sucked into the cold trap. The planet's reservoirs will slowly dry unless water can be released from the ice. Fortunately, the glaciers of Greenland and Antarctica demonstrate that there is a way for this to happen.

If water was never released from ice without melting, our own planet would look very different. The water vapour in our atmosphere would freeze at the poles and only be expelled in summer. Instead, gravity pulls piles of ice downhill to form the creeping ice sheets seen in glaciers. Such ice sheets on an eyeball planet could creep the frozen water back towards the day side, where it could melt and be vaporised once again. At the interface between ice and steam, liquid water could flow to form rivers between the dark and light planet hemispheres. This ring would be a twilight zone where life could develop with the star permanently on the horizon in a deep red sunset.

If the planet is cool enough to allow water to exist on its eyeball day side, the desert will be replaced by a sea. This sounds more habitable than a twilight strip, but there is a danger. While land and water absorb much of the radiation that reaches their surfaces, ice is very reflective and turns away the heat. Should the water ever freeze, the resultant ice would reflect heat and become even colder. This could prevent the ice from ever melting back into water.

A cool planet with exposed land might fall into this trap. The focused starlight on the day-side rocks might enhance the carbon-silicate cycle and draw too much carbon dioxide from the air. The reduction in the greenhouse gases could lower the surface temperature to below 0°C (32°F) and freeze the ocean. Now reflecting any warmth, the planet might never emerge from a permanent snowball state.

If the planet is warm enough to avoid this fate and retain liquid water, the best chance for life might be at the icy shore or under the water. This would provide a liquid pool but be out of the direct rays from the star.

A potential eyeball planet is KOI-2626-01. 'KOI' stands for *Kepler Object of Interest*, which labels a transiting planet signature discovered by the Kepler Space Telescope that has not yet been confirmed with additional observations. KOI would-be planets are denoted by numbers, rather than letters, so KOI-2626-01 is the first planet seen around the star KOI-2626. Assuming that the world does exist, KOI-2626-01 is an Earth-sized planet around a red dwarf with an orbital time of 38 days. This probably places it within the star's temperate zone, but in an eyeball-worthy tidal lock.

Given the alien climate on an eyeball planet, it is worth considering whether we should be discussing the temperate zone at all. To estimate the likelihood of supporting liquid water, the boundaries of the temperate zone assume an Earth-like planet that is uniformly distributing its heat. This is definitely not the case on a tidally locked eyeball world. Interestingly, it is possible that having a frozen and baking interface might help an otherwise Earth-like planet support water beyond the traditional limits. In the previous chapter, Gliese 581c was inside the inner edge of the temperate zone and therefore deemed likely to be a baked Venus-like world. Yet an eyeball planet's split personality might allow a runaway greenhouse effect on the day side and a cold trap on the night side. The interface would be a melted region where water could flow. This is yet another reminder that global averages do not really apply to any environment.

If the planet's atmosphere were denser than that of Earth, heat could be evenly redistributed between the day and night sides. With a day lasting longer than a year, Venus is almost in tidal lock with the Sun. Despite this, the surface temperature maintains lead-melting conditions globally over the planet. This is due to the insulation of the thick cloud cover and the strong winds in the Venusian upper atmosphere that balance out the Sun's heat. The surface of Venus is clearly not habitable, but you would die an identical burning death everywhere on the planet.

Venus's spin turns out to be an interesting conundrum. Due to its sluggish pace of 243 days per rotation, Venus spins in the opposite direction to the Earth. In the Venusian sky, the Sun rises in the west and sets in the east.

This reverse spin is very surprising. Forming within a common protoplanetary disc, the planets' orbits and spins should all be in the same direction. An anomalous rotation can often be explained by a strong collision that tilts the planet's axis. The drunken tilts of Uranus and Neptune are thought to be from major impacts late in their formation. However, the answer to Venus's reverse spin may instead lie in its atmosphere.

When bathed in sunlight, gas molecules in the Venusian atmosphere increase in speed and boost the local pressure. The pressure difference around the planet drives the hot gas into the colder region to create a patch of high-density gas.

As the gas takes time to heat, the reshuffling of the atmosphere is slightly out of sync with the Sun's motion.* Rather than the gas molecules piling into the region on the exact reverse side of the planet from the Sun, the dense patch of atmosphere ends up at an angle to the Sun's location. When the Sun's gravity pulls on this denser region, the result is a torque that turns the atmosphere. As the thick blanket of gases rolls across the planet's surface, it creates a strong enough drag to rotate the planet in the same direction.

On a shorter orbit than the Earth, Venus risks being in tidal lock with the Sun. The atmospheric drag could be preventing this from occurring, sending the planet slowly rolling in a reverse direction. Intriguingly, this mechanism to break tidal lock may be even more efficient for an Earth-like atmosphere. The thinner air will absorb less of the star's radiation, allowing more heat to permeate through to the surface-level gas. The gas dragging on the planet will therefore be more strongly affected by the temperature difference created by the star than when buried at the bottom of a thick

---

* This is also why the hottest part of the day on Earth is at about 2 p.m. rather than at noon, after the ground has had time to warm up.

Venusian atmosphere. This results in a stronger torque where it really matters, close to the planet's surface.

Without more data from planets orbiting close to their stars, it is not possible to know how many can avoid tidal lock. But should this mechanism be effective, then a planet with an Earth-like atmosphere may avoid an eyeball environment even in the temperate zone of a red dwarf.

Return to Tatooine

The blazing twin suns about Luke Skywalker's home of Tatooine supported a harsh but habitable desert landscape. But could any liquid water truly exist on a world with a circumbinary P-type orbit around two stars?

In the plane of the orbiting planet, the shape of the temperate zone around a single star is a tidy doughnut. The stellar radiation received by a planet depends on the distance to the star, resulting in a symmetrical ring where the level of heat is able to support water on an Earth-like world. Add a second star to the system and the shape of the temperate zone becomes a complex beast. The radiation a planet now receives comes from two different sources that are continually moving around one another. The temperate zone morphs into a strange, asymmetric structure that changes with time. Even if a planet remained completely stationary, the stars' motion could cause the temperate zone to move away from the planet's location, like a carpet pulled from under your feet.

Exactly how strangely shaped the temperate zone can become depends on the two stars. If their mass is very different, the radiation reaching the planet is dominated by the larger star. This can give an approximately doughnut-shaped temperate zone, but with a moving bump that follows the motion of the smaller star. If the two stars are more equally sized, the radiation will depend strongly on each of the stars' positions relative to the planet. For Sun-like or cooler stars, the temperate zone is sufficiently close to the stars to take the form of a rotating peanut. A planet within this region will feel the separate tugs of the two stars, and risks its

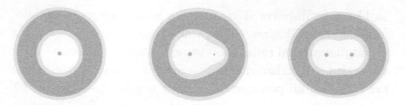

Figure 22 The temperate zone around a single star (left) and two binary star systems. The middle system has two different mass stars, while the right-hand system has two identical stars. Dark grey shows the limits of the conservative temperate zone, while the light grey indicates the extensions that allow for surface water only during the planet's early years.

orbit becoming unstable due to the stellar pair's gravity. On an unstable orbit, the dual influence of the stars will eventually scatter the planet into deep space or send it crashing into the binary. Both options bode ill for seas.

The issue of stable orbits within a wonky-shaped temperate zone ought to make it difficult to find the potentially habitable Tatooine-like worlds. Strangely, the reverse turns out to be true.

In 2015, 10 planets had been discovered in circumbinary orbits. Eight of these 10 circled their twin stars on paths very close to the limit for orbit stability. Any orbit shorter than this critical threshold would result in the planet being sent rogue or to a fiery doom. Why 80 per cent of the discovered circumbinary planets orbited so close to the stable limit is not clear. It could be a result of migration from the planets being dragged inwards by the protoplanetary disc. Any planet crossing the last stable orbit would be lost, leaving worlds that stopped migrating just inside this safety net to be observed. Alternatively, it could simply be that planets close to their stars are easier to find and these worlds are the closest possible population. Whatever the reason, it has an important consequence: with the limit for stable orbits often lying close to the temperate zone, the chances of finding circumbinary planets even within this awkwardly shaped region is higher than might be expected.

The first-discovered transiting circumbinary planet was introduced in Chapter 9. This was Kepler-16b, a Saturn-sized world that orbited twin stars every 229 days. The binary stars are both smaller than the Sun and of very unequal sizes, at 69 per cent and 20 per cent of our Sun's mass. This causes the temperate zone to be dominated by the larger star. The shape looks like a badly drawn circle, with a rotating bulge on one side due to the smaller twin.

Despite being nearly symmetrical, the temperate zone bulge makes a difference to the planet. Kepler-16b is on a circular orbit near the temperate zone's outer edge. The presence of the smaller twin forces the planet to dip in and out of the temperate zone during its year. This causes the planet's average temperature to fluctuate by about 15°C (59°F) four to five times during the orbit.[*] Of course, the local temperature away from the Equator on Earth can change by more than this during our seasons. Our summers and winters are due to the planet's tilt, tipping the northern and southern points towards or away from the Sun during the year. However, the seasonal cycle is just once per year and our global average temperature remains the same. Kepler-16b will experience fives times as many annual temperature changes that affect the whole planet.

In contrast to Kepler-16b, Kepler-453b orbits entirely within the temperate zone. The 10th discovered circumbinary world was announced in 2014 and has a radius 60 per cent larger than Neptune. Its sibling stars are also of unequal mass, with one being similar to our Sun, and the other a much smaller red dwarf with just 20 per cent of the mass. The large contrast between the stars creates a more uniformly doughnut-shaped temperate zone, dominated by the larger star. This makes it easier for Kepler-453b to orbit within its boundaries. While the planet's size ensures that this is another gaseous

---

[*] This is the temperature at the top of the atmosphere from the stars' heat. Below any gas, the temperature becomes much harder to estimate. Of course, since Kepler-16b is a gaseous world, there is no proper surface to consider.

world, it could host a rocky moon. Such a location would have Tatooine's dual stars and the rolling atmosphere of a giant gas planet in its sky.

As a small side note, film footage of the twin stars in the sky about Tatooine suggests that the temperate zone would be very lopsided. While Tatooine is portrayed as a tough environment, it is more likely that the planet would be outside the temperate zone and its surface would be uninhabitable. Sorry, Luke.

🪐

If a planet orbits just one star in a binary pair, the challenges change. In a circumstellar S-type orbit, the planet circles a single star that is moving around a stellar sibling. So how does the radiation from the sibling star affect the temperate zone of the planet's host star?

It turns out that the heat from a stellar sibling does not typically change the temperate zone boundaries. As we saw in Chapter 9, the planet usually orbits the bigger and brighter primary star in the pair. If the sibling star is distant enough that the planet's orbit is stable, then the radiation within the temperate zone is strongly dominated by the primary star.

This might imply that planets on circumstellar binary orbits are just as likely (or unlikely) to be in the temperate zone as worlds orbiting single stars. Sadly, the stellar sibling refuses to be left out of the picture. While the second source of radiation has little effect on the planet, the additional gravitational pull is a different story. Rather than allowing a planet to orbit peacefully on a circular path tucked within the temperate zone boundaries, the stellar sibling will probably pull the world on to an elongated eccentric path. With the temperate zone remaining a doughnut shape around the primary star, an eccentric orbit risks whisking the planet in and out of the most clement locations in the system.

Can a planet on an eccentric orbit ever be habitable? If the eccentricity is small and the temperate zone is wide enough, then an Earth-like planet may be able to stay within its borders. The seasons on such a world will become more extreme as the distance from the star changes, but surface water could stay liquid. If the elongated path extends beyond the temperate zone, then conditions get more tricky, but not necessarily doom-laden. A planet on an eccentric orbit moves fastest when closest to the star. The scorching summer therefore occurs during a small fraction of the year. If the hot period is sufficiently short, the planet might avoid evaporating the majority of its water before returning to cooler climes. This annual flash heating might prevent the water from completely freezing as the planet moves to its furthest point from the star. Climate calculations are heinously tricky, but have suggested that if the average radiation over the planet's year is similar to that within the temperate zone, this hot and cold balancing act on an Earth-like world could successfully retain liquid water.

Organisms developing on an eccentrically orbiting planet could adapt by hibernating during the extreme hot and cold spells of the year. This would be easiest for ocean-dwelling creatures, since a large body of water is slower to change temperature than the land. Life forms would adapt to be active when the planet was passing through the temperate zone, and would lower their metabolic rates to allow prolonged periods of inactivity and hiding when conditions were brutal.

Such protective behaviour has been exhibited by life forms on Earth. Bacteria can persevere for about a week in outer space, while microorganisms can survive in a falling meteorite if they are buried within a few centimetres of shielding rock. These experiments have only been found to work on very small creatures, but the Earth sits in the temperate zone year round. If it did not, life would have an incentive to adapt to more extreme seasons. If the variety of creatures on Earth proves anything, it is that the limits of life are very hard to guess.

The best of all possible worlds?

Our search for the ultimate habitable planet has seen us hunting the skies for a twin to the Earth. While there is no denying that the Earth is highly suited to life, is our planet truly the best possible option? Could there be planets even more likely to have developed life; the so-called *super-habitable worlds*?

The concept of super-habitability ironically begins with a land you would never want to visit. The ground is ruptured by continuous volcanic activity that fills the air with sulphur dioxide and methane. Red and yellow sulphur particles rain from the sky to strike a ground pounded by ultraviolet radiation due to the lack of a protective ozone layer. The seas run red with iron, beneath which the microbial life on the planet exists. Welcome to the Earth 2.3 billion years ago. This is the brink of the greatest extinction in our history.

About 200 million years before this moment, a blue-green bacterium appeared in the Earth's oceans. This microorganism did something never before seen on our planet; it used sunlight to take carbon dioxide and water and produce sugars and oxygen. It was the start of photosynthesis.

These tiny photosynthesising machines are known as *cyanobacteria*. The oxygen they generated initially reacted with volcanic gases to re-form carbon dioxide and water vapour, or with the iron in the water causing it to rust. But as the cyanobacteria flourished, these sinks were unable to completely absorb the output of oxygen. Oxygen flooded the atmosphere in a juncture referred to as the *Great Oxygenation Event*.

Unfortunately, most of the populations on the early Earth were anaerobic bacteria that have a toxic response to oxygen. They died in droves, wiping out a huge chunk of life on the planet. Meanwhile, the oxygen in the atmosphere reacted with methane to produce more carbon dioxide and water vapour. While both these products are greenhouse gases, neither is as effective at trapping heat as methane. The removal of methane therefore caused the temperature on the Earth to plummet, producing the massive *Huronian glaciation* that is our oldest known ice age. During this time, our planet may have become almost entirely frozen, creating a snowball Earth.

While this does not seem like an auspicious beginning for habitability, the oxidising of our atmosphere was a change key in our development. Aerobic organisms began to evolve to use the new oxygen abundance, altering the atmosphere into the air we currently breathe today. The take-home message is that time allows planets to make major modifications to their environments that can promote habitability. If we want to find a planet that probably hosts life, it is therefore possible that a super-habitable candidate will be an older world.

That said, an older world does present another challenge: the star. About 3.5 billion years ago, the Earth was in the centre of our Sun's temperate zone. As the Sun aged and brightened, the temperate zone moved outwards to leave our planet near its inner edge. Another 1.75 billion years from now and the Earth will not be in the temperate zone at all. We will have joined Venus as a lifeless desert with conditions too hot to retain surface water. A planet much older than the Earth that currently resides in the temperate zone may therefore have spent much of its youth beyond the outer boundary and not been able to develop surface life. However, this problem could be reduced if we could improve our star.

Less massive stars are slower burners than their weightier cousins, resulting in a longer life expectancy. The temperate zone around such slow-agers remains in a steady location for a longer period of time than for our Sun. However, the small and dim red dwarfs have additional problems that threaten the habitability of their planets. We have seen that worlds around these small stars risk tidal locking and being swamped in sterilising radiation, which offset any benefits from a slow-evolving temperate zone. A compromise could be an orange dwarf, a star larger than the red dwarfs but smaller than our Sun. An orange dwarf could shine for roughly double the time of the Sun, giving a significantly longer epoch for a planet to age and develop an environment suitable for life.

The star is not alone in the need to age gracefully. To maintain surface conditions, the planet must keep its geology active. In addition to radiation from the star, planets are warmed internally by the heat left over from their formation, and radioactive elements in the mantle and crust. This heat drives volcanoes and plate tectonics on the Earth, providing the rock shuffling needed for our magnetic field and carbon-silicate cycle. When our internal fires go cold, carbon dioxide will no longer be returned to the air by volcanic eruptions. Greenhouse gases will decrease and the Earth will freeze. The molten iron will also solidify in the Earth's core, and our shielding magnetic field will drop to leave our atmosphere open to stripping from the solar wind and flares.

Larger planets have a larger reservoir of internal heat to eat through, prolonging their geological life expectancy. However, this is a careful balance. If the planet's mass is too large, then the planet risks becoming a mini Neptune or at least holding on to its primitive atmosphere of hydrogen and helium. Such a thick envelope of gases would preclude habitability. Even the gravity of a massive rocky world might be too much for plate tectonics, increasing the pressure on the rocks to make movement more difficult. This would affect volcanic activity and risk our magnetic field. To maximise the chances of keeping our successful geology, the planet probably needs to be less than twice the mass of the Earth. Such a world would have a size 25 per cent larger than the Earth's with over 50 per cent more surface area.

Even with this modest increase in mass, the planet's topology would change under the stronger gravity. Our super-habitable super Earth would have a thicker atmosphere to combine with the boost in gravity and erode mountains into a flatter vista. This could change the oceans into shallow seas with long coastlines and small islands like those of the Earth's archipelagos. Such locations on our planet are rich with biodiversity, so this scenario might be excellent at fostering life.

The thicker atmosphere might also permit a different gas composition. Since all multicellular organisms need oxygen, a boost in oxygen levels on a super-habitable planet could increase the options for life. But yet again, this needs care since if our current 21 per cent oxygen content rose to 35 per cent, we would suffer from runaway wildfires. So a lift in oxygen might help life, but too much would roast it.

While a slightly different star and planet mass might lead to a super-habitable world, it at first seems obvious that we would wish to keep Earth's orbit. On a nearly circular path, the Earth avoids extreme climate fluctuations from variations in the received stellar radiation. But is this truly best for life? Life on Earth has developed assuming a constantly temperate climate. This makes the evolved creatures very sensitive to small changes. Due to the gravity of the Sun, Moon, Jupiter and Saturn, the Earth does experience subtle shifts in its orbit over tens of thousands of years. These are known as the *Milankovitch Cycles* and are like extremely long seasons. Although the change to the Earth's orbit is only a few per cent, the Milankovitch Cycle is thought to have been a major trigger for ice ages in the last few million years. If a super-habitable world were permanently on a slightly eccentric orbit, life could develop that would be capable of handling such variations more easily. This would make longer-term perturbations due to sibling planets less deadly.

Our best of all possible worlds might therefore be an old super Earth looping an orange dwarf on a slightly elliptical path. Such a planet could be our best hope for detecting Earth-like surface life. However, what if a planet develops life that does not exist on the surface? We might not be able to find signs of such life forms on a distant world, but that does not mean they do not exist.

CHAPTER FIFTEEN

Beyond the Goldilocks Zone

One of the most enticing-looking worlds in the *Star Wars* universe is the Ewok home of Endor; a forested landscape populated by hang-gliding teddybears. But Endor is not a planet. The green land is a moon orbiting the uninhabitable bulk of a gas giant.[*]

While Ewok jerky (a popular snack across the Outer Rim) may not be on the menu, habitable moons are a serious prospect. Our own gas giant planets are mobbed by moons. If such satellites could support life, the locations where ecosystems could develop would boom. But orbiting a gas giant provides more than a boost in planetary real estate. Despite being far from the temperate zone, the moons around our giants are one of the most promising places for uncovering extraterrestrial life.

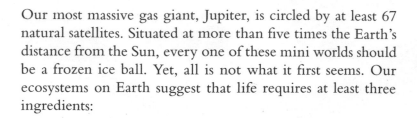

Our most massive gas giant, Jupiter, is circled by at least 67 natural satellites. Situated at more than five times the Earth's distance from the Sun, every one of these mini worlds should be a frozen ice ball. Yet, all is not what it first seems. Our ecosystems on Earth suggest that life requires at least three ingredients:

1. Biogenic elements such as carbon, oxygen and hydrogen that build living systems.
2. Water for a liquid medium in which to construct complex molecules.
3. An energy source to power life's metabolism.

---

[*] The planet's name is also Endor and it orbits a pair of binary stars, which are also called Endor. It is interesting physics, but spectacularly unimaginative nomenclature.

Situated beyond the ice line, the domain of the gas giants is packed with frozen water. However, the Sun's heat only delivers about 3 per cent of the power per unit area that we receive on Earth. To stand on the surface of Jupiter's moon, Callisto, you would need to bundle up in clothing capable of coping with average temperatures of -139°C (-218°F)

This brings us to an anomaly. While Callisto is suitably cold and still for its distant location from the Sun, its sibling moon, Io, is a fiery mass of activity. The most volcanically active place in the Solar System, temperatures on Io range from more than 1,500°C (2,700°F) to -130°C (-200°F). Admittedly, neither moon looks fit for Ewoks, but it is clear that another source of energy must be powering Io. That source is the gravitational pull from Jupiter.

While Jupiter does radiate energy by reflecting sunlight and releasing heat during the gas giant's very slow contraction, the amount is nowhere near enough to power Io. Instead, Io's generator is tidal heating. As the moon circles Jupiter, it is squished by the planet's gravitational pull just as a closely orbiting planet can be distorted by its star. Io's orbit is slightly elliptical, so the bulges raised on the moon increase and weaken as the moon moves closer and further from Jupiter. The incessant flexing raises Io's surface by more than 100m (328ft); the height of London's Big Ben. It is this stress–ball treatment that heats the moon. If Io were a lone moon, its orbit would eventually circularise and this heat source would die away. This does not happen on Io because of its large siblings.

In 1610, Italian astronomer Galileo Galilei built a telescope with which he could observe Jupiter. While tiny by today's standards, the 20-times magnification of Galileo's instrument allowed him to spot the four biggest moons around the gas giant. These became known as the Galilean moons. Galileo himself referred to the moons by a number, which was scientific if a little dull. However, the moons were simultaneously discovered by German astronomer Simon Marius. Marius named these satellites after the lovers of Zeus, the Greek mythological counterpart to the Roman god

Jupiter. Publishing in 1614, Marius referred to the four moons as Io, Europa, Ganymede and Callisto.

Moving outwards from Jupiter, the moons circle the planet in: Io, 1.8 days; Europa, 3.6 days (x 2 Io); Ganymede, 7.2 days (x 4 Io); Callisto, 16.7 days (comparatively dawdling).

Callisto is the only moon of this four whose orbit duration is not an integer multiple of the time for Io's orbit. The synchronisation of the other three forms a 1:2:4 resonance. As we saw for migrating planets in Chapter 5, resonances are difficult to break. The three moons are therefore very resistant to changing their orbits, causing the trio to all maintain elliptical paths. This allows continual tidal heating from Jupiter as it kneads the moons.

The fierce tidal heating on Io prevents liquid water from forming, but the sibling moons of Europa and Ganymede are less brutally flexed. Orbiting further from Jupiter, these moons initially appear to have the opposite problem as their cold surfaces are covered with a thick layer of solid ice. However, there is evidence that the pair are not frozen to their cores.

While the outer Jovian moon of Callisto has one of the most heavily cratered surfaces in the Solar System, the surface of Europa is one of the smoothest. Both moons should have been pounded by meteorites (in this case, crashing comets) over their lifetimes, suggesting that Europa's surface is far newer than that of its sibling. Europa's pathetic crater count points to an average age of around 65 million years, just 2 per cent of the actual age of the moon. Somehow, the moon has been resurfaced like an impossibly large public skating rink. Images of Europa's surface also reveal regions that appear to be expanding, without corresponding regions of contraction. Since the moon is unlikely to be undergoing a mad global inflation, another explanation is needed for this icy landscape.

As it happens, we see something very similar to this renewal and expansion on the Earth. Our planet's surface regularly expands during sea-floor spreading, where new crust forms as magma emerges in ridges along the ocean floor. The Earth's overall area remains constant, since similar amounts of material are subducted when one of the surface tectonic plates pushes underneath another. This comparison suggests that Europa might be the first world beyond Earth, and possibly Mercury,* to show evidence of current plate tectonics. Rather than silicate, Europa's plates are made from ice sheets. Where the icy plates pull apart, a fresh new surface is created on the moon. The mobility of the Earth's plates is evidence that the lower mantle is more plastic and moves within the Earth. Similarly, the shifting of Europa's icy surface is evidence that the moon cannot be solid ice through to its core. Instead, the frozen surface sits on a deep ocean of liquid water.

Further evidence that Europa harbours a hidden ocean comes from the appropriately named NASA space probe *Galileo*, which orbited Jupiter from 1995 to 2003, sweeping past Europa in January 2000. During this flyby, *Galileo* measured a varying magnetic field.

Europa has no magnetic field of its own, but Jupiter has the most powerful field in the Solar System. Ten times stronger than the Earth's magnetic shields, Jupiter's field originates from the blisteringly high pressures in its outer core. Crushed under the atmosphere, a metallic liquid version of hydrogen forms that acts like the Earth's molten iron centre. The flowing of this strange metal creates a current that generates a magnetic field. The inner part of Jupiter's magnetic field extends outwards to roughly 10 times the huge planet's radius, stretching to between Europa and Ganymede. The outer region is sculpted by the solar wind to form a teardrop shape, extending over 100 times the radius of the planet. As is the case with the Earth and pulsars, Jupiter's field is not perfectly aligned with the planet's rotation axis. The offset causes the

---

* NASA's *MESSENGER* space probe found evidence that Mercury may be contracting as the planet's core cools. This would also indicate that the planet's surface is tectonically active.

enveloped moons to feel a varying magnetic field strength as Jupiter rotates.

Not only can moving charged particles generate a magnetic field, but the reverse can also occur. Changes in a magnetic field will cause charged particles to move and create an electric current. The induced current will then create its own magnetic field. This effect was discovered in 1831 by the British scientist Michael Faraday. It became known as *electromagnetic induction*. Electromagnetic induction only works if charged particles can move. That is, the charges must be in a medium that can conduct a current. Metal is an excellent example of a conducting substance, but so is salt water.

When the *Galileo* probe passed Europa, it measured a fluctuating magnetic field originating from the moon. An induced magnetic field due to electromagnetic induction always opposes changes in the imposed magnetic field. As Jupiter rotated, the magnetic field strength over Europa strengthened and weakened. The moon's induced magnetic field flipped directions, attempting to weaken the strengthening field, then strengthen the weakening field.* The *Galileo* probe's measurement of this flip-flopping confirmed that Europa's magnetic field was being induced by Jupiter and not generated within the moon's core. To achieve this, the moon had to be conducting a current. Beneath its icy surface therefore had to lie an ocean of salt water.

Ice and pure water are poor conductors of electricity since they contain few mobile charged particles. Dissolve salt in the water and the situation changes. The salt separates into positively and negatively charged atoms that feel magnetic fields. As Jupiter's magnetic field waxed and waned over Europa, these salty charges flowed to create their counter field.

Due to the strength of the secondary magnetic field from Europa, this water was no briny puddle. Europa possessed a hidden global ocean where charged particles could flow freely. While it is difficult to be certain exactly how deep the

---

* Electromagnetic induction is basically fuelled by a strong dislike of change.

ocean lies, a reasonable estimate would be 10km (6mi) of ice followed by a subsurface ocean 10–100km (6–60mi) deep.

The *Galileo* probe also investigated Europa's gravitational field, which can reveal information on the internal structure of the moon. In his *Principia*, Isaac Newton showed that the gravitational force from a perfectly spherical distribution of matter was identical to a single concentration of mass at the sphere's centre. By measuring the gravitational tug from different locations close to Europa, the probe was able to record the deviations from Newton's law and recreate the true bumpy insides of the moon. These measurements suggested that Europa has an iron core surrounded by a rocky mantle beneath the deep ocean and icy lid. If the water is predominantly liquid, then Europa's oceans amount to twice the water found on Earth.

With such a copious supply of water, could Europa be teeming with hidden life? It is a possibility that is being seriously considered, with both Europe and the US planning missions to explore the icy moon further in the next decade.

Europa's hidden interior has access to water and tidal heat, but life would also require an organic starter kit. If the icy shell is thin enough to allow outside material to pass through cracks, then debris from meteorite impacts could add organic molecules to the system. A thin outer layer could even allow weak photosynthesis in the upper layers of water. High-energy particles trapped in Jupiter's strong magnetic field also bombard Europa, breaking up surface water molecules into hydrogen and oxygen. The hydrogen is too light to be retained by the moon's small gravity and escapes, leaving a supply of oxygen. This oxygen could be used in biological oxidising reactions that, similar to photosynthesis, can generate energy for organisms to survive.

If Europa's icy top is thicker, the most probable place for life is on the ocean floor. Hydrothermal vents could nurture ecosystems similar to those that flourish in the Earth's oceans. Whether these could exist depends on if Europa can summon up volcanic activity. The Jovian moon is slightly smaller than our moon, which is geologically dead. However, if the tidal heating from Jupiter is enough to melt part of Europa's rocky mantle, then hydrothermal vents could exist.

If life does exist at Europa's heart then detection will be a serious challenge. Our success would depend on the mixing of the more complex telltale organic molecules between the ocean bottom and icy shell. If hints of hidden life can be found in the ice, then we might be able to deduce what lies out of sight.

Should life be found on Europa, it is likely to be completely independent of any life on Earth. While the Earth and Mars could possibly have exchanged early microbial material via meteorites tossed between the planets, Europa's distance makes such sharing far more improbable. Finding life on the Jovian moon would therefore tell us a great deal about how easy it was for life to begin.

Stepping out further from Europa takes you to the biggest moon in our Solar System, Ganymede. The third Galilean moon is a little larger than the planet Mercury, but only half as massive due to having nearly 50 per cent of its mass in ice.

Unlike Europa, Ganymede generates it own magnetic field without electromagnetic induction from Jupiter. The moon probably creates the field similarly to the Earth within a molten iron core. This also produces an aurora on Ganymede as the magnetic shielding channels charged particles towards the moon's poles. With an icy surface similar to Europa's, Ganymede has a tenuous atmosphere of oxygen from split-water molecules. Collisions with oxygen atoms would illuminate a splendid red aurora if you were to look up from a spot on the moon's surface. It is the aurora that provides a clue to the moon's internal structure.

Still under the umbrella of the outer extent of Jupiter's magnetic field, Ganymede experiences both its own field and the fluctuating strength emanating from the rotating giant planet. As Jupiter's magnetic field waxes and wanes over the moon, Ganymede's aurora is tugged back and forth. The expected result was calculated to be a change of about 6 degrees, but observations by the Hubble Space Telescope show a smaller 2-degree shift. The difference can be explained by an additional Europa-like induced magnetic field within Ganymede, repelling the effect of Jupiter's imposed variations. The existence of a second magnetic field on the moon is proof that Ganymede also has a subsurface ocean. Although this situation was suspected by the *Galileo* probe, the results were far less conclusive than for Europa. The aurora observations by the Hubble Space Telescope nailed down the speculation.

Could Ganymede's hidden ocean be habitable as well? It turns out that the situation is less promising than for Europa. Three times more massive than its inner sibling icy moon, the pressures near Ganymede's core are considerably higher, risking the freezing of the lower ocean into a thick icy layer. The result would make Ganymede a deep-ocean water world like Gliese 1214b, with the silicate sea floor sealed off from the water. Further from Jupiter than Io or Europa, Ganymede also has less tidal heating. Its surface ice is much older than that on Europa, dating back to several billion years, and with no evidence of recent geological activity such as plate tectonics. This suggests that the ocean may be deep under the surface ice, around 150–300km (90–185mi) down. The combination indicates that Ganymede's waters may be both unable to benefit from surface organics or sunlight, and also be blocked from hydrothermal vents.

The third and final moon to harbour a possible ocean in the Jovian system is Callisto. The outermost Galilean moon, Callisto is not in resonance with its three large siblings. It therefore has to forego tidal heating and rely on the dwindling heat from its formation to warm its heart. Its icy surface is the oldest in the Solar System, with a heavily cratered visage that denies the presence of geological activity. It was therefore suspected that Callisto, at least, had to be an entirely frozen world. The moon's internal heat should have been insufficient to keep its interior from freezing solid.

Despite this expectation, the *Galileo* probe revealed a surprise. Like Europa and Ganymede, Callisto has an induced magnetic field caused by Jupiter. This pointed to another subsurface salty ocean. It seems that the moon's icy crust is better at retaining Callisto's limited heat supply than had previously been suspected. That said, with a solid surface and small supply of energy, Callisto is the least likely of the three icy siblings to support life.

The tiny, shiny moon

While the hidden oceans of the Jovian moons had to be deduced from the subtle shifts of their interiors, Saturn's moon Enceladus had no such modesty. The moon was observed rocketing 250kg (550lb) of water vapour into space every second through cracks in its icy shell. Such watery expulsions are known as *cryovolcanoes*, which spew ice and water rather than lava. The plumes stretch 500km (310mi) above the surface at the south pole, making the tiny moon the smallest volcanic body in the Solar System.

Like Jupiter, Saturn is not short of natural satellites. At least 62 encircle our second-largest gas giant, ranging in size from asteroid-sized moonlets only a kilometre across, to a moon approaching Ganymede in size. Smaller than the moonlets are the rings of Saturn, which comprise microscopic dust through to bodies measuring a few hundred metres. These extend over thousands of kilometres in a disc only about 10m (33ft) thick. Saturn's major moons have sizes between 10 and 150 per cent of our own Moon, and all sit beyond the main rings.

In Greek mythology, the personifications of the Earth and Sky had two races of divine children known as the Titans and the Giants. The leader of the Titans was Kronus, who later bore the Roman name Saturn. He and his consort (and slightly disturbingly, also his sister) Rhea (Roman name: Ops), birthed the gods Zeus (Jupiter), Poseidon (Neptune) and Hades (Pluto).[*]

---

[*] Roman mythology adopted most of the Greek gods but switched to Latin names. The Titans and Giants appear predominantly in the Greek mythology.

The naming of Saturn's major moons was proposed by the British polymath John Herschel in 1847, when he suggested that the large satellites take the names of the Titans and Giants who were the siblings of Saturn. Enceladus was discovered by John Herschel's father, William Herschel, and was named after one of the Giants.

The sixth-largest moon of Saturn, Enceladus is tiny and very shiny. At only 500km (310mi) across, the moon is just over a seventh of the size of our Moon, and can almost fit within the borders of England or Arizona.

The moon orbits Saturn within its diffuse and outermost ring, which is fed particles from Enceladus's geysers. The moon's shiny surface is due to the continual resurfacing of highly reflective fresh ice from the spouting water, making the moon one of the most reflective bodies in the Solar System. Mirroring back even the small amount of radiation received near Saturn results in a particularly cold surface, with a noon average surface temperature of -198°C (-324°F).

Until the joint US and European Cassini-Huygens missions arrived at Saturn in 2004, very little was known about Enceladus. Being so small and further than even Jupiter's moons from the Sun, it was assumed to be a dead and frozen world. The discovery of geysers revealed both the presence of water and the geological activity needed to drive the plumes. Enceladus clearly required a rethink.

It was initially thought that Enceladus's subsurface water might be confined to the active geysers around the moon's south pole. This region showed a *tiger-stripe* pattern of cracks where the water had burst through the ice. Then further observations picked up a slight wobble in Enceladus's orbit that could most easily be explained by the sloshing of a global ocean about 26–31km (16–19mi) deep. Such a wobble is similar to that seen when spinning a raw egg.* The water depth is about 10 times deeper than the average ocean depth on Earth.

---

* Do not try this near the edge of a table.

With water easily accessible as it spurted from Enceladus's interior, the *Cassini* orbiter grabbed a sample as it flew through the plumes. It found a melange of water, carbon dioxide, methane, salt and ammonia crystals. Combined with the heat source that drives the moon's geysers, this mix of organic compounds could create an environment for life.

Far from the Sun, Enceladus is heated with the same tidal flexing that fires up the Jovian moon trio. The tiny Saturn moon is in resonance with its sibling, Dione (named after a Titan in Greek mythology). Enceladus orbits Saturn twice in the time it takes Dione to circle once. Like the Galilean inner moons, tugs from Dione keep the orbit of Enceladus slightly elliptical so that Saturn's grip tightens and weakens during its orbit. However, this does not entirely explain Enceladus's power source. The expected heating from Saturn is lower than the energy that seems to drive the powerful geysers. Possibly the deficit is compensated by a residue of the internal heat generated during the moon's formation, or left over from an era when Enceladus might have been on a more elliptical orbit.

The easy access to Enceladus's water makes it a tempting prospect for exploring life on these icy moons. Unlike on Europa, a visiting probe would not need to land or drill through kilometres of ice to analyse the ocean's contents. This is offset by the longer distance to the Saturn system, making the moon a seriously long hike from Earth. The Cassini-Huygens mission took seven years to reach Saturn, compared with the 2016 arrival of *Juno* at Jupiter, which took five years. Current planned missions are therefore focused on Europa, but Enceladus remains a tantalising target for future searches for non-Earth life.

The moon with liquid lakes

Enceladus may be small but Saturn's moon, Titan, more than makes up for its dearth. While 62 moons have been recorded orbiting the ringed gas giant, Titan comprises more than 96 per cent of their combined mass. The second most massive

Saturn satellite is Rhea, which has less than 2 per cent of Titan's mass and a third of the size. This makes Titan the second-largest moon in the Solar System, being topped by a mere 2 per cent in size by Ganymede.

Like the watery ice moons, Titan has a slightly elliptical orbit that results in tidal heating from Saturn. Unlike the other moons, the origin of this eccentricity is not clear. Titan has no sibling moon large enough to keep its motion perturbed, and Saturn's pull should have dragged the moon into a circular orbit. It is possible that Titan suffered a relatively recent collision and not enough time has passed to circularise its path around the gas giant.

Regardless of the origin, Titan's eccentricity causes Saturn to flex the moon during its 16-day orbit. The distortions to the moon's shape were measured by the *Cassini* probe and found to be far higher than expected for a solid rocky body. Titan's surface bulged by 10m (33ft) rather than the anticipated 1m (3ft). For comparison, the Sun and our Moon cause the Earth's crust to rise by about 50cm (20in) and our open oceans to rise about 60cm (24in). Titan's squishiness suggests that this moon also harbours a subsurface ocean.

Due to the size of the moon, the pressure near Titan's core makes it likely to be encased in a layer of frozen ice. As is the case with Ganymede and exoplanet water worlds, deep-sea life is therefore significantly less likely. However, the surface of Titan is completely different from the surfaces of the icy moons.

Rather than a top lid of ice covered by a thin envelope of gas, Titan has a thick atmosphere with a surface pressure 50 per cent higher than there is on Earth. This makes the moon one of four rocky worlds in our Solar System with significant atmospheres. Venus wins for the densest atmosphere, while the Earth and Mars both have thinner atmospheres than Titan. But as is the case with our planetary neighbours, Titan's air is far from breathable.

Titan's thick atmosphere was first perceived as early as 1908. The moon had been discovered on 25 March 1655 by the Dutch astronomer and physicist Christiaan Huygens. He and his older brother, Constantine Huygens, Jr, were interested in the manufacture of scientific instruments and had observed the moon using one of their own designs. Around 250 years later, Catalan astronomer José Comas Solà measured a change in the brightness from the centre of the moon's surface to its perceived edge. He interpreted this gradient as the presence of an atmosphere.

In the 1940s, Kuiper (of Kuiper belt fame) examined the wavelengths of light being absorbed by Titan's atmosphere. He concluded that the air contained methane but was not certain if this was the dominant gas. This was resolved during a flyby by the two NASA *Voyager* probes in 1980 and 1981. The pair confirmed that a thick atmosphere surrounded the moon, which obscured the view to the planet's surface. The gases found were about 95 per cent nitrogen with up to 5 per cent methane. Interactions with the limited ultraviolet sunlight that reached the moon allowed the methane to form more complex molecules of hydrogen and carbon, such as ethane.* These heavier hydrocarbons then sank to the ground as solids or liquids, creating the orange haze that blocks the clear view of the moon's surface. Collectively, these sinking hydrocarbons are known as *tholins*, after the Greek for *sepia ink*, due to their reddish–brown colour.

As on Venus and Earth, Titan's atmosphere provides the moon with a greenhouse cloak to boost the surface temperature by about 10°C (50°F). This fails to protect the moon against freezing temperatures since due to its incredibly distant location it receives just 1 per cent of the sunlight that reaches Earth. Surface temperatures on Titan are therefore a staggeringly cold –180°C (–290°F). Such temperatures preclude the possibility of liquid water. Instead, liquid methane and ethane pool in lakes on Titan's surface.

At a (rather warm compared to Titan) temperature of 0.01°C (32.02°F), water can exist as a solid, liquid and gas. This strange temperature is known as the *triple point* of water.

---

* Methane is CH_4 (four hydrogen atoms and one carbon). Ethane is a longer chain of C_2H_8 (eight hydrogen atoms and two carbon).

That temperatures on the Earth are close to this triple point is what allows abundant quantities of ice, water and vapour to exist on our planet's surface. Together, these phases give us the water cycle whereby water moves from clouds to rain to ice and snow. While temperatures on Titan are far from water's triple point, they are close to that of methane. This can exist in all three phases at -182°C (296°F), and forms a *methane cycle* of clouds, rain and lakes on Titan.

Quite why Titan gathered a thick atmosphere when the similarly sized Ganymede and Callisto failed to is not entirely clear. One possibility is that comet impacts on the Jovian moons were faster and harder due to Jupiter's stronger gravity. These major collisions could have stripped the gases from the larger Jovian moons. Alternatively, the colder environment around the more distant Saturn was more effective at trapping gases within ices during moon formation, allowing these to later vaporise into the atmosphere.

Given the more familiar environment of surface lakes and rivers, could life exist on Titan's methane shores? The answer depends on whether methane could replace water as a medium for biological reactions. This is highly speculative and further hindered by the colder temperatures on Titan making it more difficult to dissolve the necessary organics into solution.

Evidence that Titan is indeed a barren world came from the Cassini–Huygens mission. While Titan's smoggy gases scatter optical light and prevent a view of the surface, the longer infrared wavelengths can pass through to the ground. The *Cassini* probe's instruments used this to map the moon from above. Meanwhile, the *Huygens* lander dropped down to Titan's surface. The lander separated from *Cassini* on Christmas Day in 2004 to begin a three-week descent through the smoggy atmosphere.

Unlike on Mars, where the thin atmosphere is too thin to slow down an incoming spacecraft to a safe landing speed, the *Huygens* probe was able to touch down using parachutes. It landed in a cloud of hydrocarbon dust on 14 January 2005, to become the first probe to land on a world in the outer Solar System.

The *Huygens* lander was equipped with three hours of battery life, which was used during the final stages of the descent and 72 minutes on the moon's surface. The lander sent back around 80 images from the surface, none of which showed any signs of movement or plant life. Assuming that life would spread to all regions of the moon as on Earth, Titan would seem to be devoid of macroscopic life.

While our Solar System's outer moons may harbour life, their ecosystems are hidden below thick icy or opaque gases. Similar moons around extrasolar planets would appear lifeless rocks in observations from Earth. But what about giant planets that have migrated closer to their stars? Of the plethora of gas planets found in the temperate zone, could one of these host a more Earth-like moon?

CHAPTER SIXTEEN

The Moon Factory

'Moons are where planets were in the 1990s,' German astrophysicist René Heller explains, citing the date just prior to the first exoplanet discoveries. 'We're on the brink.'

Heller is obsessed with moons. Specifically, his research focuses on the formation and possible detection of moons around extrasolar planets. It is on these *exomoons* that Heller believes we might find life.

With distances too great for a robotic drop-in, habitats on exomoons must create an imprint in the atmosphere to be detectable. Target moons therefore cannot harbour hidden ecosystems under a frozen lid, but must have surface life. Titan aside, the most likely location for landscapes desirably riddled with bacteria or Ewoks is within the temperate zone. This precludes moons like the Solar System's icy satellites, but what if Jupiter had migrated inwards to the Earth's position? Would the icy lids of Europa, Ganymede and Callisto have melted to leave three potentially habitable surfaces?

Such terrestrial moons may even have advantages over a planet. A major issue for planets orbiting close to the cooler red dwarf stars is tidal locking. With one face permanently turned to the star, the temperature on the planet can become a bimodal split of inhospitable scorching heat and freezing darkness. Tidal locking on a moon would be to the planet, rather than the star. A normal day and night cycle could therefore roll over a moon even within a red dwarf's temperate zone.

We have so far discovered five times as many gas giants in the temperate zone compared with terrestrial-sized worlds. If these Jupiter analogues are as loaded with moons as our own giants, then exomoons are the main surface territory in this region. This seems like a great prospect for

habitability, but can an Earth-like moon such as the Ewoks'
Endor really exist?

To support its abundance of trees and furry teddybears, Endor
must have an atmosphere. Given the thick envelope of gases
surrounding Titan, it might seem a cinch for a moon to hold
on to the requisite supply of air. However, Titan has one
advantage over a moon inside the temperate zone: it sits in
very cold space.

Gas molecules are lost from the upper atmosphere if they
can move fast enough to escape the planet's (or moon's)
gravity. As the density of gas thins near the atmosphere edge,
gas with enough speed is unlikely to encounter any obstacles
before it escapes the planet altogether. To keep an atmosphere
around for a long time, such loss needs to be very slow. This
can be achieved in two ways: either the planet needs to be
very massive to create a strong gravitational pull, or the
atmosphere molecules need to be moving very slowly. Moons
in our outer Solar System are not very large, but the tops of
their atmospheres are very cold. This makes it difficult for the
gas molecules to gain enough speed to leave a moon.

However, move the moon inwards towards the Sun and
the atmosphere will warm to allow gas to escape. The
temperature in the upper atmosphere of the Earth is a hundred
times hotter than for Titan. In such a temperate region,
Titan's gravity would be too weak to hold on to its own
atmosphere.

If the moon had an Earth-like atmosphere, the loss of
oxygen could potentially be replaced by photosynthesis and
lost carbon dioxide from silicate weathering. However, losing
nitrogen from the atmosphere of a potentially habitable world
would be devastating for life. Nitrogen is an element that is
difficult to engage in chemical reactions, and nitrogen
molecules are a buffer that reduces the risk of forest fires. It is
also a key component in proteins and DNA used by all life on
Earth. Lose the nitrogen, and the forest moon of Endor is

dead. A habitable moon therefore needs to be massive enough for its gravity to hold on to this warmer atmosphere for billions of years.

For a rocky world to retain nitrogen and oxygen at similar upper atmosphere temperatures to the Earth for over 4.6 billion years (the life of our Solar System), its mass needs to be slightly larger than that of Mars. This brings us to a problem: the largest moon around our largest gas giant is just 23 per cent the size of Mars. Even if Ganymede sat in the Sun's temperate zone, it would be a world with no atmosphere. So can we make moons larger than Ganymede? The answer depends on how the moons are born.

🪐

The moons in our Solar System were not formed using one uniform mechanism, but by at least three independent methods. Our own Moon is at the weirder end of the natural satellite population. With 1.2 per cent (or 1/81th) of the Earth's mass, our Moon has a particularly large mass compared with its planet. It is outdone only by Pluto and its giant moon, Charon, which is nearly 12 per cent (1/9th) of Pluto's mass. Conversely, the moons of Mars, Jupiter, Saturn, Uranus and Neptune are less than 0.025 per cent (1/4,000th) of their planet's mass. So how did small worlds like the Earth and Pluto acquire such large satellites?

The answer is that both our Moon and Charon were formed during a giant impact. Our Moon is thought to have been created when the young Earth was struck by an object the size of Mars, throwing material from the impactor and the Earth's surface into orbit around our planet. This coalesced into our Moon, producing a composition similar to the Earth's mantle, but devoid of lighter elements that were vaporised in the impact and escaped into space.

Another outlier in the moon population is Neptune's moon Triton. With a diameter of 2,700km (1,680mi) and a mass 40 per cent larger than Pluto, Triton is the largest of Neptune's 14 known moons and the seventh largest in the Solar System.

Unlike the other large moons, Triton circles Neptune in the opposite direction to the planet's spin. This strange retrograde orbit suggests that Triton and Neptune did not form together, but that Triton was a passer-by that was captured into orbit by Neptune's gravity. Triton's composition is very similar to that of Pluto, indicating that this moon may have formed in the Kuiper belt as a dwarf planet, but was later snagged by Neptune during the giant planet's migration.

Capturing a moon as large at Triton is a considerable achievement. To be locked into orbit, the moon to be must be moving slowly enough that it cannot escape the planet's gravity. However, on a typical orbit around the Sun, the dwarf planet Triton would be moving too swiftly to be captured by Neptune. Triton's large mass also prevents a collision with an existing Neptunian moon providing a sufficient braking force without a crash so catastrophic that the colliding worlds would be destroyed.

A popular theory is that Triton was once a binary dwarf planet, similar to Pluto and Charon. Rotating about their centre of mass as the binary orbited the Sun, the speed of each sibling compared with Neptune would be alternately slightly faster and slower than the net binary speed.

As Neptune approached the binary, the planet's strong gravity began to dominate over the paired dwarf planets and ripped them apart.* The dwarf planet currently moving slightly slower than the binary speed was captured by the giant planet, while its twin shot off into space.

The captured dwarf planet subsequently became Triton. The new giant moon scattered and collided with the smaller existing satellites of Neptune, to leave Triton as the overwhelmingly dominant moon, comprising 99.5 per cent of the mass that orbits around Neptune.

---

* In more scientific language, Neptune's strengthening gravity caused the binary's Hill radius to shrink until it was smaller than the separation between the two dwarf planets. At that point, they were no longer bound to one another and the binary was broken.

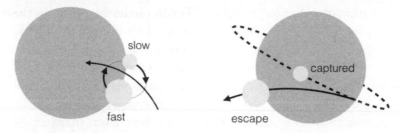

Figure 23 Neptune's moon Triton may have been part of a binary of dwarf planets. Orbiting around its sibling, each dwarf planet alternates moving with, and against, the motion of the binary's centre of mass. Each in turn, therefore, moves slightly faster then slower with respect to Neptune. When Neptune's gravity disrupts the binary, the slower dwarf planet is captured to become a moon.

The majority of the moons in the Solar System did not form through giant impact or capture. Instead, they formed in a disc of gas and dust that circled the young gas giants. These circumplanetary discs are similar to scaled-down versions of the protoplanetary disc that circled the young Sun. Like the protoplanetary disc, the circumplanetary disc is formed from gas whose circular motion balances the gravitational pull from the giant planet.

Yet, the two discs are not entirely analogous. In particular, the dust within the circumplanetary disc can feel the gravitational tug from both the planet and the star. These two gravitational pulls result in a specific region where moons can form in a circumplanetary disc. Too close to the planet, and the tidal forces from the planet's gravity will rip apart any forming moon. Too far away, and the star's tug will lure the moon on to an unstable orbit. The strict outer limit is the planet's Hill radius, beyond which the star's gravity is a stronger force than that of the planet. In practice, the moon needs to form within about a third of the Hill radius to ensure that it stays safely tied to its parent world. The inner edge for moon formation is where the planet's gravity dominates inside the moon's structure and rips it to pieces. This is the same limit that we saw for chthonian hot Jupiters whose atmospheres overspilled when they approached too close to the star.

As the main method for building a moon, the prospects for habitability depend on how large a satellite forming in a circumplanetary disc can grow. If circumplanetary discs are larger around big planets, then the plus-sized Jupiter worlds seen orbiting other stars offer a promising location. So is it possible to build a giant Mars-sized moon that could hold on to an atmosphere?

Like in a protoplanetary disc, a key to super-sizing a moon is the availability of solid ice. If water can freeze, the amount of material for moon building around the biggest planets can allow a Mars-sized moon to form. But the necessity for ice runs us into a problem; moons that form inside the temperate zone are within the ice line. Temperatures here are too warm for water to freeze in the circumplanetary dust. A super-sized moon therefore needs to be built beyond the ice line and migrate inwards, or be captured like Triton. Since capture is difficult, potentially habitable moons could be water worlds whose icy content has melted as they travelled into the temperate zone. As we saw in Chapter 14, this does not preclude life but it does make conditions different from those on Earth.

Of course, forming beyond the ice line is not a guarantee of ice. The volcanic Jovian moon of Io is proof that moons may have very different temperatures from their planets.

Within a circumplanetary disc, a proto-moon will feel two tugs of gravity from the star and planet and also two sources of heat. The reflected heat and thermal radiation from the planet creates the circumplanetary disc's own ice line. Closer to the planet than this ice line, the disc receives too much heat from the planet for ice to solidify. Moons forming in this region will be dry, regardless of how far the planet orbits from the star. A potentially habitable moon therefore needs to form outside two ice lines: further from the star than the protoplanetary ice line and further from the planet than the circumplanetary ice line.

Once the moon has water, it then needs to keep it. As the Galilean moons, Enceladus and Titan demonstrate, moons of gas giants are very susceptible to tidal heating. This provides an additional inner radius around the planet that a moon on an eccentric orbit must not cross if it is to hold on to any lakes and oceans. If the moon's orbit takes it within this line, then even a moon squarely in the star's temperate zone risks being squeezed into a runaway greenhouse state. This line is akin to an additional temperate zone edge for moons.[*]

The exact location of the circumplanetary temperate edge depends on the sizes of both the planet and the moon. Tidal heating is stronger for more massive planets and more massive moons. Assuming a slightly eccentric orbit to trigger tidal heating, a Mars-sized moon can therefore orbit closer to a planet than a moon the size of the Earth, without having a Venus-like meltdown. Similarly, potentially habitable moons around a planet more massive than Jupiter will need to be further out than those around a world the size of Neptune.

If the moon is on a circular orbit, then it will be immune to the flexing that produces tidal heating. A potentially habitable moon can then pass the circumplanetary temperate edge until the planet's heat evaporates any surface water. In this case, a larger moon turns out to be able to radiate more energy per area than a smaller satellite can, allowing it to keep cool more easily. While tidal heating overwhelms this effect, without this flexing larger worlds are the cooler option.

The additional heating is not always a disadvantage for life. An added boost from tidal heating may make a moon just outside the star's temperate zone able to maintain surface lakes. If Mars had been a moon rather than a planet, it might have been habitable. This feature could be particularly

---

[*] This is also referred to as the 'habitable edge', a term we will tweak to the 'temperate edge' for the same reasons as we avoid using 'habitable zone'.

useful for the large number of planets that are on eccentric orbits around their stars. If an elongated orbit swung outside the stellar temperate zone, the planet could provide a boost in heat to its moons that would prevent surface water from freezing.

Tidal heating could also counter the second plague of small worlds: that of stifled geological activity. To maintain a carbon silicate cycle or stir up a magnetic field, the moon needs a supply of internal heat to roll around tectonic plates and power volcanic eruptions. Without this, a moon or planet may tumble into a snowball state even within the temperate zone. The Earth's internal heat comes from its collisional formation and radioactive rocks. A smaller world would have a smaller heat store and its geological engines would halt much earlier. It is hard to know exactly how much mass is needed to keep geological action over the age of the solar system, but an estimate would be about 25 per cent of the Earth's mass. This presents a problem for a Mars-sized moon, which would have only 10 per cent of the Earth's mass – that is unless tidal heating can come to the rescue.

As the most volcanically active location in the Solar System and with just 1.5 per cent of the Earth's mass, Io is proof of the power of tidal heating. A more moderate dose could help to keep the moon's surface moving while avoiding sending the atmosphere into a runaway greenhouse.

Even with this helping hand, a smaller moon might still struggle to power a magnetic field as strong as that around the Earth. But could the planet provide further assistance? Jupiter generates the strongest magnetic field in the Solar System, creating a barrier between its inner moons and the solar wind. If a moon sits under the magnetic umbrella of its giant planet, can it avoid having its atmosphere stripped by the star?

It turns out that this protection is not the metaphorical free lunch. While the planet's magnetic field diverts high-energy particles, these particles can then become trapped in radiation belts. The Earth has at least two such regions, known as the

Van Allen belts after their discovery by American space scientist James Van Allen. These doughnut-shaped bands of high-energy particles circle the Earth and pose a serious threat to any man-made satellites in that region. A gas giant planet with a magnetic field thousands of times stronger than the Earth's will have correspondingly more dangerous radiation belts that could sterilise a moon. Indeed, space missions to Jupiter and its inner moons have to take care that the high levels of radiation do not take the spacecraft out of action. Whether the surface of a moon will be safe for life will depend on the orientation of the planet's magnetic field and the moon's orbit.

The creation of our habitable moon therefore needs to begin outside both the star and the circumplanetary ice line. Once large enough to maintain an atmosphere, the moon must migrate inwards with its planet to the star's temperate zone. The surface temperature on the moon needs to be high enough to melt ice into lakes, but not so warm that an inhospitable deep-ocean world is formed. Whether a temperate climate can be maintained will then depend on the moon's orbit. If the moon successfully dodges the radiation belts and is massaged only gently by the planet, then a long-standing geological cycle may begin. This could be the beginnings of Endor.

With our own gas giants in the frozen outer reaches of our Solar System, the process of making a moon with a habitable surface is speculative. What we need to confirm these theories is to find an exomoon.

Searching for Endor

To date, no moons have been discovered orbiting planets outside our Solar System. However, if René Heller's prediction is correct, we may be on the cusp of a flood of discoveries. But what is the best way to hunt for exomoons?

Given the challenges in finding a tiny planet orbiting a far larger and brighter star, it is not surprising that finding an even smaller moon is difficult. A number of approaches are being tried to sniff out evidence of accompanying moons, and one of these we have met before.

The planets transiting across Kepler-138 were introduced in Chapter 6 as their notably small mass had been measured during a search for exomoons. This hunt looked for tiny variations in the time between transit appearances caused by the pull of an orbiting moon. In this case, the small differences in the orbit of Kepler-138d were due to the pull of its neighbour, Kepler-138c. While not an exomoon discovery, this was a successful demonstration of how a moon's influence could be spotted.

There is also the chance that a moon could be seen as an additional drop in light as the planet transits across the star. If the moon and planet are well separated, the moon could cause an independent dip as it blocks light from the star. Alternatively, the moon and planet may combine to produce a single dip, but the total amount of light obscured drops as the moon moves from beside the planet to overlapping it in front or behind.

One project to search for these tiny fluctuations in brightness and transit crossings is called HEK, standing for the *Hunt for Exomoons with Kepler*. Led by David Kipping (the researcher who examined the planets of Kepler-138), HEK studies transiting planets found by the Kepler Space Telescope for hints of hidden moons. Due to the telescope's sensitivity, a moon the mass of Mars is unlikely to be found, but a moon roughly twice the mass of Mars might be detectable. Such a big moon could be difficult to form in a circumplanetary disc, but could be a captured satellite like Triton. While HEK has not yet found an exomoon, the careful examination of the Kepler data has revealed a number of hidden planets as a particularly good consolation prize.

Surprisingly, another possible technique for detecting exomoons may be direct imaging. Given the difficulty of spotting the dim heat signature of a planet compared with the roaring inferno of a star, a moon seems like an impossible prospect. Yet, this is not necessarily true. Additional heating from the planet could allow moons to remain hot and luminous for a long time, even if they are far from the star. The newest generation of telescopes could be able to spot these anomalous heat signatures in the infrared.

The combination of moons and rocky worlds that we are now starting to find offers many places where life could occur. But how would we know if we do have alien neighbours?

CHAPTER SEVENTEEN

The Search for Life

‘How long do you think it will be before we find life on another planet?’

I was asked this question by Michel Mayor, the researcher who discovered 51 Pegasi b and set in motion the science within these pages. Mayor's achievements had just bagged him the prestigious 2015 Kyoto Prize in Basic Sciences. The award had been marked by a day of exoplanet talks, ending in a coffee break during which Mayor had graciously listened to my fumbled self-introduction. His question revealed that even the pioneers of the exoplanet field are still in the dark when it comes to the search for life.

As we have begun to find smaller planets in potentially temperate orbits around their stars, our curiosity about their likeness to the Earth has steadily mounted. Could any of these planets not only be habitable, but actually be inhabited? The first step is to consider what kinds of life we are likely to detect. Given the remarkably varied conditions in which microbial life can survive, extraterrestrial microbes are expected to be far more common than intelligent life. But how do we go about detecting something that cannot wave back at us?

‘Strongly suggestive’

Imagine that we have found a rocky, Earth-sized planet, which is orbiting a star in its temperate zone. The planet is too distant for a probe to visit it within our lifetime, but we are able to examine its atmosphere as it transits across its star. How can we tell if life has developed on this world?

An ideal test would be to observe a world we know harbours life and look for giveaway evidence. Somewhat serendipitously, the *Galileo* spacecraft gained this opportunity en route to Jupiter. The original plan for the NASA Jupiter mission was to launch the probe from the cargo bay of the space shuttle *Atlantis*. Using a powerful booster rocket, the *Galileo* spacecraft could have taken a direct path to the outer Solar System. However, the tragic destruction of the *Challenger* space shuttle in 1986 led to strict new safety regulations that forbade the fully fuelled rocket booster from travelling inside the space shuttle before launch. *Galileo* was ultimately launched from *Atlantis* using a much lower-powered propulsion in 1989.

To enable the spacecraft to still reach Jupiter, *Galileo* took a route that brought it close to Venus and the Earth. These close passes give the probe a helping kick from the planet's gravity, acting like a planet-booster rocket. Such slingshots are known as *gravitational assists* and are popular manoeuvres to reduce the amount of fuel needed to reach an interplanetary destination.

The *Galileo* spacecraft did not waste this close approach to Earth. Turning its sensors our way, the spacecraft was able to observe our planet from space. If we could not conclude that the Earth was heaving with life from a mere 1,000km (620mi) away, then there was little hope of uncovering the secrets of worlds light years distant. But what exactly were we looking for?

Life is not a passive addition to a planet. Processes such as photosynthesis, respiration and decay change the composition of an atmosphere. What the *Galileo* probe needed to find was something all life – from intelligent to microbial – produced in large enough quantities that it could be detected either in an atmosphere, or by a measurable change in the global property of the planet. In other words, *Galileo* was hunting for a *biosignature*.

By looking at the fingerprint of absorbed wavelengths as sunlight streamed through our atmosphere, the instruments on *Galileo* picked out the molecules in our air. As two of the by-products of life, oxygen and methane were clearly present. Water was also in the atmosphere, and the light reflecting from the planet's surface suggested that it could be found as the solid, liquid and gas needed for a water cycle. *Galileo* additionally detected structured radio signals that were a flag for intelligent life. But without the radio detection, would the data have been sufficient to be certain life existed? The resulting paper in *Nature* in 1993 was led by US astronomer and science communicator Carl Sagan. The tone from Sagan and the scientific team was cautious, declaring the signs from *Galileo* 'strongly suggestive of life on Earth'.

Based on Earth, oxygen might seem like the ideal biosignature. Since the appearance of cyanobacteria 2.5 billion years ago, our atmosphere has been infused with oxygen from photosynthesising plants, algae and bacteria. This led to the development of more complex, multicellular organisms that require a sizeable quantity of oxygen molecules. Without life, the easily reacting oxygen would have been mopped up by rusting the iron-rich waters and creating carbon dioxide and water from volcanic gases. It takes the continual renewal of oxygen from biological sources to keep the level at 21 per cent in our atmosphere.

With our searches for surface life focusing on worlds in the temperate zone, starlight will be a major source of energy for the planet. It is very reasonable to assume that life will select to make use of this source, developing organisms that photosynthesise. While there are forms of photosynthesis that do not produce oxygen, at least on Earth oxygen is the dominant by-product. This makes it a good guess that the atmospheres of inhabited worlds will be oxygen rich.

The catch is that life is not the only source of oxygen. In the atmosphere, ultraviolet radiation from the star can break apart water molecules into hydrogen and oxygen. Like in the tenuous atmospheres of the Jovian ice moons, the light hydrogen can escape the planet to leave an oxygen-rich air. This process could be particularly successful for the planets orbiting red dwarfs; the easiest worlds to examine due to the proximity of the temperate zone to the dim star. If the planet is geologically inactive without the regular production of gases that react with oxygen to remove it from the air, a fake oxygen biosignature could be created.

So if oxygen cannot be guaranteed to signpost life, what about methane? Nearly all methane on Earth is produced by microbes that live in the guts of animals, the bogs and marshes of wetlands, and help decompose the remains of plants. Like oxygen, methane also needs to be continually replenished to remain in our atmosphere. Without a regular supply, methane will react with oxygen to produce water and carbon dioxide, or be split by ultraviolet sunlight to re-form as the more complex ethane. The Earth's current supply of methane would disappear from the air within 10 years if it were not topped up. If methane is detected in a planet's atmosphere, it therefore must be regularly added.

But there is a known problem with using methane as a biosignature: Titan. While methane is continually lost from Titan due to the incoming sunlight, the moon has a subsurface stash of icy methane that can be liberated into the atmosphere in cryovolcanoes. This has allowed methane in Titan's atmosphere to persist for billions of years. The Earth also has a few non-biological sources of methane, including volcanoes and the deep-sea black smoker vents. Sufficient geological action could therefore infuse methane into the atmosphere of a dead world.

Potentially, biological and abiotic sources of methane can be distinguished. A methane molecule is made up of a carbon atom and four hydrogen atoms. Life on Earth prefers using carbon-12, the most common form of carbon with six neutrons in its nucleus. The alternative choice is carbon-13, whose extra neutron creates a heavier atom that needs more energy to react. Inorganic material on Earth has 89.9 times more carbon-12 than carbon-13, but in living matter this is enhanced to 95. The difference may not be huge, but it can be detected. On Titan, the *Huygens* probe measured the ratio between the two carbon types in the atmospheric methane as 82.3, much closer to the inorganic value on Earth. If we could manage to differentiate between the different carbon atoms in methane on an exoplanet, we might have a flag for life.

A more promising biosignature might be a combination of molecules. When oxygen and methane are present in the atmosphere together, they combine to form carbon dioxide. If both are seen in substantial quantities, it implies that there is a source that is continually resupplying the atmosphere with the two molecules to keep it out of equilibrium. Such combined biosignatures are more difficult to fake abiotically but not impossible. A Titan-like moon orbiting an oxygen-rich planet may be seen as a single world with a mixed atmosphere of methane and oxygen.

An extra hint that the atmosphere was being influenced by life would be changes during the planet's year. As winter rolls around to spring, the biosphere on Earth undergoes a substantial alteration when plants bloom in the warmer weather. This can be seen clearly in the levels of carbon dioxide in our atmosphere. Measurements taken from the Mauna Loa Observatory in Hawaii since 1958 show a steady rise in carbon dioxide from global climate change, but also an annual wiggle from seasonal fluctuations.

With the increase in sunlight as spring begins, photo-synthesising plants grow fresh leaves to pull more carbon dioxide from the air. When winter closes in, the foliage dies off and the carbon dioxide level rises. Of course, spring on one half of the Earth corresponds to winter in the other. A global change therefore requires an unequal quantity of plant-covered land in the northern and southern hemispheres. The Earth has more land plants on its northern continents than on its southern ones, so the annual fluctuations in carbon dioxide match the spring and winter seasons on the northern side of the planet. A perfectly even distribution of foliage on any planet is pretty unlikely, and the detection of such a cycle as a planet orbits its star would be hard to explain without a biological presence.

Green bump, red edge

The atmosphere is not the only planet property affected by life. Another potential biosignature is the planet's colour or – to put it more scientifically – the wavelengths of radiation the planet predominately reflects. We see the Earth's land as green because the widespread photosynthesising plant life reflects green light. This is from the chlorophyll in the plant cells that scatters wavelengths around 500nm,* but absorbs incoming light at slightly shorter and longer wavelengths. Our plants also strongly reject wavelengths longer than our eyes can detect. In the infrared starting at around 700–800nm, radiation is either reflected or passes through the plant untouched. It is a boundary known as the *vegetation red edge*. This reflection causes plants to show up brightly on infrared cameras and has been used by satellites to spot vegetation cover.

These two features give the light from our planet a distinctive green bump and red edge. If the wavelengths reflected from an extrasolar planet could be observed, similar peaks and troughs might form the biosignature fingerprint for plant life.

But are the wavelengths our plants absorb ubiquitous to all possible vegetation on alien worlds? To make an educated guess, we need to first understand exactly why our plants are green.

At first it seems that reflecting green light is a strange choice for plant evolution. While the Sun appears yellow from the Earth's surface, it actually emits most of its energy at green wavelengths. This becomes a yellow hue when blue light is scattered by the atmosphere. So by reflecting green light, our plants have evolved to reject the largest chunk of the Sun's power.

However, a closer look suggests a reason for this choice. The energy the Sun emits at a particular wavelength is a combination of that wavelength's energy and its intensity or brightness. If we think of radiation as made up of bite-sized photons, the energy emitted depends on the energy of each photon and the number of photons.

---

* 1nm = 0.000000001m

The Sun actually emits more red photons than green, but the longer wavelength of red light means that each red photon carries less energy than a green photon. On the other hand, blue photons carry more energy but are far less numerous. Plants might have adapted to take advantage of the high quantity of red photons and the individual power of blue photons. This left less use for green photons that are neither as numerous as red photons nor individually as powerful as blue photons.

If plants adapt to use the wavelengths of photons that are high energy or highly numerous, then their properties will depend on both the atmosphere and the star. The atmosphere will rob the planet's surface of any wavelengths it absorbs, while the temperature of the star will determine how much energy is emitted at each wavelength. We can see the effect of a similar type of filtering on the Earth. Plants growing in the ocean and beneath layers of sand are not the same colour as surface plants. These take on different hues as they adapt to absorb the wavelengths that reach them. For example, water allows through blue light but absorbs red, stifling the evolution of plants below a certain depth that use red photons. The result is algae in a range of browns, reds and purples that reflect the photons they have not developed mechanisms to use.

In contrast to cherry picking the best photons for photo-synthesis, the rejection of wavelengths beyond the vegetation red edge may be to stop the plants from becoming too hot. If infrared wavelengths as well as those of visible light were absorbed, the energy overload might irreversibly damage proteins and fry the plants. This gives us an explanation for why our planet throws off infrared and green light into space. So what would happen if we changed our Sun?

As we previously noted, it is not unreasonable to suspect that life-supporting planets within the temperate zone would be covered with light-harvesting organisms. However, they may not be green.

Around the cool red dwarf stars, most energy is emitted in the infrared. Photosynthesis is still thought to be possible at these longer wavelengths, but the lower energy of the photons may mean that twice as many are needed. In order to absorb as much light as possible, plants on a planet around a red dwarf may be black, rather than green. Moreover, with the infrared photons now needed, such a planet's vegetation may not have the defining red edge, or it may be shifted to a different wavelength.

Plants on planets around more luminous stars than our Sun may experience the opposite problem as the surface is bombarded with high-energy blue photons. The foliage may develop a blue tint to reflect these photons and to prevent them from being fried. In any eventuality, the expression 'the grass is always greener' is unlikely to apply to alien worlds.

As with a biological fingerprint in the atmosphere, an indication that a planet's colour is linked with foliage would be a seasonal variation. This could differentiate between growing foliage and a false signature from rocky minerals. Jupiter's moon Io starts reflecting radiation strongly at the blue wavelength of 450nm. Light redder than this bounces from the surface, while anything blue-wards is absorbed. This is not due to any form of plant life miraculously surviving on the highly volcanic world, but to a sulphur coating spewed over the moon during the copious eruptions.

The biosignature capable of confirming life on a planet around another star is unlikely to be a single measurement, but a combination of the ways in which life changes a planet. As for a complex jigsaw puzzle, we will need to find a large number of the pieces before we can be certain that the end image is one of an inhabited world.

So how long does Michel Mayor believe it will take before we can suggest that a planet could be inhabited? 'Twenty-five years,' he told me. 'It'll take a generation.'

Glossary

astronomical units (au) Measure of distance for planetary systems. 1 au (= 150,000,000km/0.000016 light years) is the distance between the Sun and the Earth.

binary stars Two stars that orbit around their common centre of mass.

carbon–silicate cycle The geological process that regulates the quantity of carbon dioxide in the air on Earth. As carbon dioxide is a greenhouse gas and can trap heat, this acts as a thermostat for our planet.

centre of mass Balance point in a system of objects (stars, planets, moons and so on), where the pulls from their gravity cancel one another.

eccentricity Measure of how elongated a planet's orbit is. Zero per cent eccentricity is a perfect circle.

exomoon Moon orbiting an exoplanet.

exoplanet Planet orbiting a star that is not our Sun.

gas giant planet Planet like Jupiter, Saturn, Uranus and Neptune that has a solid core engulfed in a huge atmosphere, thousands of kilometres thick.

greenhouse effect Ability of gases in a planet's atmosphere to absorb and reflect infrared radiation, warming the planet.

Hill radius Distance around a planet (or other object), where its own gravity dominates over that from the star. Smaller objects (such as planetesimals) in this region are pulled towards the planet.

hot Jupiter Gas giant planet that orbits close to the star.

ice line Distance from a star where it becomes cold enough for ice to form in the protoplanetary disc. Also called the snow line, or frost line.

inclination Angle a planet's orbit is tilted from the plane of the other planets in the system (or the plane perpendicular to the rotation axis of the star).

isolation mass The mass of a growing planet once it has eaten all planetesimals within its orbit.

Kepler Space Telescope NASA space observatory that detects extrasolar planets by the transit technique.

Kozai–Lidov Mechanism Ability of another body (like a binary star or another planet) to affect the eccentricity and inclination of a planet's orbit.

light year Distance light travels in one year. It is equivalent to 63,240au, or 9,500,000,000,000km (6,000,000,000,000mi).

migration Changing of a planet's orbit, usually in the direction of the star. Type I and II migrations refer to orbit changes due to the drag of the gas in the protoplanetary disc. Planetesimal-driven migration is from the scattering of smaller planetesimals.

planetesimals Asteroid-sized rocks a few kilometres to a few hundred kilometres in size that formed on the way to building planets.

protoplanetary disc The disc of gas and dust that circles newly forming stars, out of which planets are formed.

radial velocity technique Detection of a planet from the slight wobble in a star's motion. This gives the orbital time for the planet (and therefore its distance from the star), and the minimum estimate of its mass.

red dwarf Star smaller and cooler than our Sun. Also called an M dwarf.

resonant orbits The orbits of neighbouring planets with periods (time for one orbit) in exact integer ratios; for example, one planet orbits twice in the time taken for the second planet to orbit once. Such orbits are very stable and difficult to break.

roguo planct A planet that does not orbit a star.

super Earth Planet with a radius of 1.25–4 Earth radii. Such planets may be rocky or have Neptune-like thick atmospheres.

Spitzer Space Telescope NASA space observatory that detects infrared heat.

temperate zone Region around a star where surface temperatures on a planet exactly like the Earth allow liquid water to exist. Also called the habitable zone, or Goldilocks zone.

terrestrial planet A planet like Mercury, Venus, Earth and Mars that is predominantly rock with a thin atmosphere.

tidal heating Heat generated by flexing a planet or moon due to the varying gravitational pull from the star or planet when in an eccentric orbit.

tidal locking Orbit of a planet or moon where one side always faces its host (star or planet).

transit technique Detection of a planet from the slight dip in a star's light as the planet passes across its surface. This gives the orbital time for the planet and its radius.

transit timing variations (TTV) Changes in the time between planet transits due to the gravitational tugs of other planets (or potentially moons) in the same system. This gives the mass of the planets.

Further Reading

Listing all the research papers whose results shaped the understanding of planets in *The Planet Factory* would make for a list as long as the book. To avoid such inundation, I've tried to pick out original sources or reviews for a few key results that are tricky to track down.

Preface: The Blind Planet Hunters

The discovery of the first planet found around a Sun-like star. 51 Pegasi b: M. Mayor & D. Queloz 1995. A Jupiter-mass companion to a solar-type star. *Nature* 378:355–359.

The first transiting exoplanet discovery, HD 209458. The two papers announcing the find were published in the same January 2000 edition of the journal, which actually came out in December 1999: 1. D. Charbonneau *et al.* 2000. Detection of planetary transits across a Sun-like star. *The Astrophysical Journal Letters* 529:L45–48; 2. G. Henry *et al.* 2000. A transiting '51 Peg-like' planet. *The Astrophysical Journal Letters* 529:L41–44.

Chapter 2: The Record-breaking Building Project

A comprehensive review of the research on how to build a planet from dust to planetesimals: A. Johansen *et al.* 2014. The multifaceted planetesimal formation process. In *Protostars and Planets VI* (University of Arizona Press, Tuscon, USA, 2014). This review accompanies talks presented at the *Protostars and Planets VI* meeting, which are freely available online: www.mpia.de/homes/ppvi.

Chapter 4: Air and Sea

Fred Whipple's review of Ernst Öpik's work: F. Whipple 1972. Ernst Öpik's research on comets. *Irish Astronomical Journal Supplement* 10:71–76.

Chapter 5: The Impossible Planet

For excellent descriptions of new exoplanet discoveries, Sean Raymond's blog, *PlanetPlanet* (planetplanet.net), is a great resource.

Proposal of Jupiter's grand tack: K. Walsh 2011. A low mass for Mars from Jupiter's early gas-driven migration. *Nature* 475:206–209.

The Nice Model: R. Gomes *et al.* 2005. Origin of the cataclysmic Late Heavy Bombardment period of terrestrial planets. *Nature* 435:466–469.

The Nice Model II: H. Levison *et al.* 2011. Late orbital instabilities in the outer planets induced by interaction with a self-gravitating planetesimal disk. *The Astronomical Journal* 142:152–162.

The planet with the density of polystyrene, WASP-17b: D. Anderson *et al.* 2010. WASP-17b: An ultra-low density planet in a probable retrograde orbit. *The Astrophysical Journal* 709:159–167. The discovery was described in *Wired* (where Coel Hellier is quoted) 2009: Aack, no breaks! Giant new exoplanet goes the wrong way, http://bit.ly/2kuEaGc.

Chapter 6: We Are Not normal

A precise mass measurement for Kepler-93b was finally announced by C. Dressing *et al.* 2015. The Mass of Kepler-93b and the composition of terrestrial planets. *The Astrophysical Journal* 800:135–141

The mass measurement for Kepler-138d (then named KOI-314c) by transit timing variations: D. Kipping *et al.* 2014. The hunt for exomoons with Kepler (HEK): IV. A search for moons around eight M dwarfs. *The Astrophysical Journal* 784:28–41. The press release by the Harvard-Smithsonian Center for Astrophysics (including a quote from Kipping) 2014: Newfound planet is Earth-mass but gassy, http://bit.ly/2kvR47c.

The rough rule of thumb that suggests planets larger than 1.5 Earth radii are mini-Neptunes rather than rocky planets: L. Rogers 2015. Most 1.6 Earth-radius planets are not rocky. *The Astrophysical Journal* 801:41–53.

Investigations of whether super Earths form from a different-shaped protoplanetary disc: 1. H. Schlichting 2014. Formation of close in super Earths and mini-Neptunes: required disk masses and their implications. *The Astrophysical Journal Letters* 795:L15–19; 2. S. Raymond & C. Cossou 2014. No universal minimum-mass extrasolar nebula: evidence against in situ accretion of systems of hot super Earths. *Monthly Notices of the Royal Astronomical Society: Letters* 440:L11–15.

A hot Jupiter's atmosphere overflowing to leave a mini Neptune: F. Valsecchi, F. Rasio & J. Steffen 2014. From hot Jupiters to super Earths via Roche lobe overflow. *The Astrophysical Journal Letters* 793:L3–8.

The hot Jupiter-broom for piling up material to create a super Earth: S. Raymond, A. Mandell & S. Sigurdsson 2006. Exotic Earths: forming habitable worlds with giant planet migration. *Science* 313:1413–1416.

The discovery of the Kepler-11 with six planets was described (with quotes of amazement from Jack Lissauer) by NASA 2011: NASA's Kepler Spacecraft discovers extraordinary new planetary system, http://go.nasa.gov/2kKtimo and a number of other sites, including the *Guardian* 2011: NASA scientists discover planetary system, http://bit.ly/2lv7ydU.

Super Earth formation at the edge of the dead zone: S. Chatterjee & J. Tan 2014. Inside-out planet formation. *The Astrophysical Journal* 780:53–64.

Computer modelling of how migration might change direction: C. Cossou *et al.* 2014. Hot super Earths and giant planet cores from different migration histories. *Astronomy & Astrophysics* 569:A56–71.

Chapter 7: Water, Diamonds or Lava? The Planet Recipe Nobody Knew

The models of planetesimal formation around a carbon-rich star discussed by Torrence Johnson and Jonathan Lunine: T. Johnson *et al.* 2012. Planetesimal compositions in exoplanet systems. *The Astrophysical Journal* 757:192–202. Johnson's joke about 'no snow beyond the snow line' and Lunine's observation on carbon worlds was in an accompanying news release by the Jet Propulsion Laboratory 2013: Carbon Worlds May be Waterless, Finds NASA Study, http://go.nasa.gov/2kVk0WA.

Possible changes in the geology of rocky planets with different compositions: 1. C. Unterborn et al. 2014. The role of carbon in extrasolar planetary geodynamics and habitability. The Astrophysical Journal 793:124–123; 2. J. Bond, D. O'Brien & D. Lauretta 2010. The compositional diversity of extrasolar terrestrial planets, I, In situ simulations. The Astrophysical Journal 715:1050–1070.

Measuring the carbon abundance in 55 Cancri: J. Teske et al. 2013. Carbon and oxygen abundances in cool metal-rich exoplanet hosts: A case study of the C/O ratio of 55 Cancri. The Astrophysical Journal 778:132–140.

A protoplanetary disc with C/O > 0.65 might still spawn carbon-rich planets: J. Moriarty, N. Madhusudhan & D. Fischer 2014. Chemistry in an evolving protoplanetary disc: Effects on terrestrial planet composition. The Astrophysical Journal 787:81–90.

Could 55 Cancri e be a carbon world? N. Madhusudhan, K. Lee & O. Mousis 2012. A possible carbon-rich interior in super Earth 55 Cancri e. The Astrophysical Journal Letters 759:L40–44.

Cambridge University news release on 55 Cancri e (including quote from Madhusudhan) 2015: Astronomers find first evidence of changing conditions on a super Earth, http://bit.ly/1c0gsu1.

The geology of the potentially magnesium rich Tau Ceti planets: M. Pagano et al. 2015. The chemical composition of τ Ceti and possible effects on terrestrial planets. The Astrophysical Journal 803:90–95.

The temperature variations on 55 Cancri e: 1. B.-O. Demory et al. 2016. Variability in the super Earth 55 Cnc e. Monthly Notices of the Royal Astronomical Society 455:2018–2027; 2. B.-O. Demory et al. 2016. A map of the large day-night temperature gradient of a super Earth exoplanet. Nature 532:207–209.

The fractionating column atmosphere of CoRoT-7b: L. Schaefer & B. Fegley 2009. Chemistry of silicate atmosphere of evaporating super Earths. The Astrophysical Journal Letters 703:L113–117. The accompanying article by Washington University in St Louis (including a quote from Fegley) 2009: Forecast for discovered exoplanet: clouds with a chance of pebbles, http://bit.ly/2ku8GQF.

The helium atmosphere of Gliese 436b: R. Hu, S. Seager & Y. Yung 2015. Helium atmosphere on warm Neptune- and sub-Neptune-sized exoplanets and applications to GJ 436b. *The Astrophysical Journal* 807:8–21. The accompanying news release by the Jet Propulsion Laboratory (with a quote from Seager) 2015: Helium-shrouded planets may be common in our Galaxy, http://go.nasa.gov/2k5MrNG.

Chapter 8: Worlds Around Dead Stars

A wonderful account of the pulsar planet discoveries is given in Ken Croswell's *Planet Quest: the Epic Discovery of Alien Solar Systems* (Free Press, New York, USA, 1997).

For a lively and readable work on pulsars themselves, try Geoff McNamara's *Clocks in the Sky: the Story of Pulsars* (Praxis Publishing Ltd, Chichester, UK, 2008).

The first millisecond pulsar discovery: D. Backer *et al.* 1982. A millisecond pulsar. *Nature* 300:615–618.

Wolszczan and Frail's discovery is also described in an article by Charles DuBois in *Penn State News* 1997: Planets from the Very Start, http://bit.ly/2kurW0x.

Alex Wolszczan's first-hand account of the pulsar planet discoveries: A. Wolszczan 2012. Discovery of pulsar planets. *New Astronomy Reviews* 56:2–8.

The signature flash of black widow pulsar PSR J1311-3430: H. Pletsch *et al.* 2012. Binary millisecond pulsar discovery via Gamma-ray pulsations. *Science* 338:1314–1317.

The star that became a diamond world orbiting pulsar PSR J1719-1438: M. Bailes *et al.* 2011. Transformation of a star into a planet in a millisecond pulsar binary. *Science* 333:1717–1720.

Chapter 9: The Lands of Two Suns

Walker's first-hand account of nearly discovering a planet around γ Cephei: G. Walker 2012. The first high-precision radial velocity search for extra-solar planets. *New Astronomy Reviews* 56:9–15.

The planet around γ Cephei was finally announced in: A. Hatzes *et al.* 2003. A planetary companion to γ Cephei A. *The Astrophysical Journal* 599:1383–1394.

The survey of discs around young stars in Taurus-Auriga: R. Harris *et al.* 2012. A resolved census of millimeter emission from Taurus multiple star systems. *The Astrophysical Journal* 751:115–134.

Comparison of planets in binary systems with different separations: J. Wang *et al.* 2014. Influence of stellar multiplicity on planet formation. II. Planets are less common in multiple-star systems with separations smaller than 1500au. *The Astrophysical Journal* 791:111–126.

A research review of how a binary star can disrupt the planet-building process for circumstellar orbits: Thébault & Haghighipour 2014. Planet formation in binaries. In *Planetary Exploration and Science: Recent Advances and Applications* (Springer Geophysics, Heidelberg, Germany, 2015).

Models to determine if the protoplanetary disc around γ Cephei would have sufficient mass to form a gas giant: H. Jang-Condell, M. Mugrauer & T. Schmidt 2008. Disk truncation and planet formation in γ Cephei. *The Astrophysical Journal Letters* 683:L191–194.

The planet detection around Alpha Centauri B: X. Dumusque *et al.* 2012. An Earth-mass planet orbiting a Centauri B. *Nature* 491:207–211.

The fresh analysis of the data that called the planet into question: A. Hatzes 2013. The radial velocity detection of Earth-mass planets in the presence of activity noise: The case of α Centauri Bb. *The Astrophysical Journal* 770:133–148.

The announcement of the Tatooine world, Kepler-16b: L. Doyle *et al.* 2011. Kepler-16: A transiting circumbinary planet. *Science* 333:1602–1606.

The theory for how the pulsar, white dwarf and gas giant triplet (PSR 1620-26) came to exist was proposed about 10 years after the discovery: S. Sigurdsson *et al.* 2003. A young white dwarf companion to pulsar B1620-26: Evidence for early planet formation. *Science* 301:193–196.

Whether the observed variations in the binary transits such as NN Serpentis really imply the presence of planets is discussed by J. Horner *et al.* 2012. A detailed investigation of the proposed NN Serpentis planetary system. *Monthly Notices of the Royal Astronomical Society* 425:749–756.

The planet in a three-star system, HD 131399Ab: K. Wagner *et al.* 2016. Direct imaging discovery of a Jovian exoplanet within a triple-star system. *Science* 353:673–678.

Phil Plait's Slate article on HD 131399Ab 2016: An alien planet orbits in a triple-star system... and we have photos, http:// slate.me/29JnqoY.

Chapter 10: The Planetary Crime Scene

Mike Brown's blog is a great read for posts on the outer Solar System: www.mikebrownsplanets.com.

The discovery of the dwarf planet Sedna: M. Brown, C. Trujillo & D. Rabinowitz 2004. Discovery of a candidate Inner Oort Cloud planetoid. *The Astrophysical Journal* 671:645–649.

Changes in young Neptune's orbit that may have scattered the distant dwarf planets: R. Dawson & R. Murray-Clay 2012. Neptune's wild days: Constraints from the eccentricity distribution of the classical Kuiper Belt *The Astrophysical Journal* 750:43–71.

Measuring our Solar System's centre of mass using pulsar signals: N. Zakamska & S. Tremain 2005. Constraints on the acceleration of the solar system from high-precision timing. *The Astrophysical Journal* 130:1939–1950.

Massive eccentric planets can be dragged on to circular orbits by the gas disc: B. Bromley & S. Kenyon 2014, The fate of scattered planets. *The Astrophysical Journal* 796:141–149.

The planet with one of the 'fiercest storms in the Galaxy', HD 80606b: G. Laughlin *et al.* 2009. Rapid heating of the atmosphere of an extrasolar planet. *Nature* 457:562–564. The press release by NASA where Laughlin is quoted 2009: Spitzer watches wild weather on a star-skimming planet, http://go.nasa.gov/2ltA3J6.

The ejection of a planet from υ Andromedae A to explain the highly perturbed orbits of two of the other planets: E. Ford, V. Lystad & F. Rasio 2005. Planet–planet scattering in the υ Andromedae system. *Nature* 434:873–876.

Smaller planets have less eccentric orbits: V. Van Eylen & S. Albrecht 2015. *The Astrophysical Journal* 808:126–145.

Chapter 11: Going Rogue

Sean Raymond's excellent piece in *Aeon* magazine: Life in the dark, http://bit.ly/2jF2R2g.

Did our Solar System once have an extra gas giant planet?: D. Nesvorny & A. Morbidelli 2012. Statistical study of the early Solar System's instability with four, five and six giant planets. *The Astronomical Journal* 144:117–136.

The discovery of the incredibly distance world, HD 106906b, with its debris disc: V. Bailey *et al.* 2014. HD 106906 b: A planetary-mass companion outside a massive debris disk. *The Astrophysical Journal Letters* 740:L4–9.

The follow-up observations that identified the disc asymmetry: P. Kalas *et al.* 2015. Direct imaging of an asymmetric debris disk in the HD 106906 planetary system. *The Astrophysical Journal* 814:32–43.

Observations of tiny dense clouds that could collapse to planet-sized objects (the same team also coined the term 'globulettes'): G. Gahm *et al.* 2013. Mass and motion of globulettes in the Rosette Nebula. *Astronomy & Astrophysics* 555:A57–73.

The potential for a rogue Earth to maintain heat has been considered in a few different publications, including: 1. D. Stevenson 1999. Life-sustaining planets in interstellar space? *Nature* 400:32; 2. G. Laughlin & F. Adams 2000. The frozen Earth: binary scattering events and the fate of the Solar System. *Icarus* 145:614–627; 3. D. Abbot & E. Switzer 2011. The steppenwolf: a proposal for a habitable planet in interstellar space. *The Astrophysical Journal Letters* 735:L27–30; 4. J. Debes & S. Sigurdsson 2007. The survival rate of ejected terrestrial planets with moons. *The Astrophysical Journal Letters* 668:L167–170.

Chapter 12: The Goldilocks Criteria

The boundaries of the temperate zone (also known as the habitable or Goldilocks Zone): J. Kasting, D. Whitmire & R. Reynolds 1993. Habitable zones around main sequence stars. *Icarus* 101:108–128.

The Venus Zone: S. Kane, R. Kopparapu & S. Domagal-Goldman 2014. On the frequency of potential Venus analogs from Kepler data. *The Astrophysical Journal Letters* 794:L5–9.

Chapter 13: The Search for Another Earth

The discovery of the first transiting planet in the temperate zone, Kepler-22b: W. Borucki *et al.* 2012. Kepler-22b: A 2.4 Earth-radius planet in the habitable zone of a Sun-like star. *The Astrophysical Journal* 745:120–135. News release by NASA (with quote by Borucki) 2011: NASA's Kepler mission confirms its first planet in the habitable zone of a Sun-like star, http://go.nasa.gov/2kpfix8.

The discovery of Gliese 581c (declared 'the most Earth-like of all known exoplanets' at the time): S. Udry *et al.* 2007. The HARPS search for southern extra-solar planets XI. Super Earths (5 and 8 M⊕) in a 3-planet system. *Astronomy & Astrophysics Letters* 469:L43–L47.

The existence of Gliese 581d and g was questioned in P. Robertson *et al.* 2014. Stellar activity masquerading as planets in the habitable zone of the M dwarf Gliese 581. *Science* 345:440–444.

The Earth-sized Kepler-186f: E. Quintana *et al.* 2014. An Earth-sized planet in the habitable zone of a cool star. *Science* 344:277–280.

Natalie Batalha's description of trying to find the transit of a planet with a size and orbit similar to our own was on an Advexon TV NOVA documentary 2014: Kepler 186f – Life after Earth, http://bit.ly/1xPw9Jj.

The frequency of Earth-sized worlds: 1. F. Fressin *et al.* 2013. The false positive rate of Kepler and the occurrence of planets. *The Astrophysical Journal* 766:81–100; 2. C. Dressing & D. Charbonneau 2013. The occurrence rate of small planets around small stars. *The Astrophysical Journal* 767:95–114.

The discovery of our nearest exoplanet: G. Anglada-Escudé *et al.* 2016. A terrestrial planet candidate in a temperate orbit around Proxima Centauri. *Nature* 536:437–440.

Chapter 14: Alien Vistas

The Hubble Space Telescope's attempt to explore Gliese 1214b's atmosphere: L. Kreidberg *et al.* 2014. Clouds in the atmosphere of the super Earth exoplanet GJ1214b. *Nature* 505:69–72.

Could water be stored in the mantle to prevent a water world? N. Cowan & D. Abbott 2014. Water cycling between ocean and mantle: super Earths need not be water worlds. *The Astrophysical Journal* 781:27–33.

The failure of oceans to regulate the planet's temperature: D. Kitzmann *et al.* 2015. The unstable CO_2 feedback cycle on ocean planets. *Monthly Notices of the Royal Astronomical Society* 452:3752–3758.

Life on a gas giant core: R. Luger *et al.* 2015. Habitable evaporated cores: Transforming mini-Neptunes into super Earths in the habitable zones of M dwarfs. *Astrobiology* 15:57–88.

Sean Raymond has a great piece for *Nautilus* magazine: Forget 'Earth-Like' – we'll first find aliens on eyeball planets. http://bit.ly/1vRsb1J.

The ability of an eyeball world to support an atmosphere: M. Joshi, R. Haberle & R. Reynolds 1997. Simulations of the atmospheres of synchronously rotating terrestrial planets orbiting M dwarfs: Conditions for atmospheric collapse and the implications for habitability. *Icarus* 129:450–465.

Climate and water content of eyeball planets: R. Pierrehumbert 2011. A palette of climates for Gliese 581g. *The Astrophysical Journal Letters* 726:L8–12.

The turning of the atmosphere could break tidal lock: J. Leconte *et al.* 2015. Asynchronous rotation of Earth-mass planets in the habitable zone of lower-mass stars. *Science* 347:632–635.

The shapes of temperate zones around binary stars and stable circumbinary orbits: S. Kane & N. Hinkel 2013. On the habitable zones of circumbinary planetary systems. *The Astrophysical Journal* 762:7–14.

The temperate zone boundaries in Figure 22 were sketched based on calculations from the website described in T. Müller & N. Haghighipour 2014. Calculating the habitable zone of multiple star systems with a new interactive website. *The Astrophysical Journal* 782:26–43. http://astro.twam.info/hz.

The influence of the second star for planets in circumstellar binary systems: S. Eggl *et al.* 2012. An analytics method to determine habitable zones for S-type planetary orbits in binary star systems. *The Astrophysical Journal* 752:74–84.

The possibility of maintaining liquid water and life on an Earth-like world in an eccentric orbit: 1. D. Williams & D. Pollard 2002. Earth-like worlds on eccentric orbits: excursions beyond the habitable zone. *International Journal of Astrobiology* 1:61–69; 2. S. Kane & D. Gelino 2012. The habitable zone and extreme planetary orbits. *Astrobiology* 12:940–945.

A super-habitable world: 1. René Heller's 2015 article for *Scientific American* 312:20–27. Better than Earth; 2. R. Heller & J. Armstrong 2013. Superhabitable worlds. *Astrobiology* 14:50–66.

Chapter 15: Beyond the Goldilocks Zone

Evidence for plate tectonics on Europa: S. Kattenhorn & L. Prockter 2014. Evidence for subduction in the ice shell of Europa. *Nature Geoscience* 7:762–767.

Chapter 16: The Moon Factory

Our outer Solar System's moons and moon formation research review: R. Heller *et al.* 2014. Formation, habitability and detection of extrasolar moons. *Astrobiology* 14:798–835.

The formation of Triton as a destroyed binary: C. Agnor & D. Hamilton 2006. Neptune's capture of its moon Triton in a binary-planet gravitational encounter. *Nature* 441:192–194.

Chapter 17: The Search for Life

Hunting for biosignatures on Earth: C. Sagan *et al.* 1993. A search for life on Earth from the Galileo spacecraft. *Nature* 365:715–721.

The ratio of carbon-12 to carbon-13 in Titan's atmosphere measured by the *Huygens* probe: H. Riemann *at al.* 2005. The abundances of constituents of Titan's atmosphere from the GCMS instrument on the Huygens probe. *Nature* 438:779–784.

Nancy Kiang's 2008 article for *Scientific American* 298:48–55. The colour of plants on other worlds.

Finally, if you feel ready for a few equations in a very readable text, I recommend Caleb Scharf's *Extrasolar Planets and Astrobiology* (University Science Books, Sausalito, CA, USA, 2009).

Acknowledgements

First, I would like to offer a blanket apology to anyone who was forced to listen to a planet fact when they had just wanted me to pass the toast at breakfast (although I stand by the analogy with carbon worlds). To those of you listed here, I can honestly say that this would never have been a problem without your merciless encouragement. You have only yourselves to blame.

This book would never have begun if it were not for Jim Martin and Anna MacDiarmid at Bloomsbury. Thank you for bearing with me while I learned how to assemble something longer than a few thousand words.

I owe a huge debt to my technical readers – René Heller, Dimitri Veras, Mordecai-Mark Mac Low, Jonti Horner, Sourav Chatterjee, Shogo Tachibana, Caleb Scharf, Sean Raymond, Johanna Teske and Eric Ford – for giving up their free time to read through my chapters. Without your supportive feedback, I would have stuffed this book under a mattress. Most especially, I'd like to thank Stephen Kane, who supported this project from day one, from discussions for the book plan through to last-minute panicked queries.

I have been truly fortunate to have had excellent mentors throughout my career, without whom I could never have written a book. My secondary school English teacher, Pat Huntzinger, looked past my atrocious dyslexic spelling to tell me I could write, and even read through my early attempts as a novelist (everyone died at least once: it was more gory than a holiday on 55 Cancri e). My graduate thesis advisor, Greg Bryan, has been a never-ending source of encouragement over the last decade, as has my postdoctoral advisor, James Wadsley. Ralph Pudritz at the Origins Institute at McMaster University turned my interest in planets into the unquenchable thirst that has only been partially sated by writing an entire volume on the subject.

I also need to thank Kelly Roy Komura, who created a real baby while I constructed a book baby, but never told me I had it easy when my manuscript didn't need a nappy change. Also all my friends who never once doubted I would finish this book and supported my struggles both in person and online (and who once suggested that I contribute a planet fact in the middle of a wedding ceremony – you know who you are).

Most of all, thanks to my parents for being my best friends. To my dad for his positive plans of action each time world-ending chaos seemed inevitable in my life, and to my mum for leading by example in writing her own PhD thesis and for being my very first editor while I was in school (an utterly thankless task). I love you both.

Index

Page numbers in **bold** refer to glossary entries.